Visual Communication for Cybersecurity

RIVER PUBLISHERS SERIES IN SECURITY AND DIGITAL FORENSICS

Series Editors:

WILLIAM J. BUCHANAN
Edinburgh Napier University, UK

ANAND R. PRASAD
NEC, Japan

Indexing: All books published in this series are submitted to the Web of Science Book Citation Index (BkCI), to SCOPUS, to CrossRef and to Google Scholar for evaluation and indexing.

The "River Publishers Series in Security and Digital Forensics" is a series of comprehensive academic and professional books which focus on the theory and applications of Cyber Security, including Data Security, Mobile and Network Security, Cryptography and Digital Forensics. Topics in Prevention and Threat Management are also included in the scope of the book series, as are general business Standards in this domain.

Books published in the series include research monographs, edited volumes, handbooks and textbooks. The books provide professionals, researchers, educators, and advanced students in the field with an invaluable insight into the latest research and developments.

Topics covered in the series include, but are by no means restricted to the following:

- Cyber Security
- Digital Forensics
- Cryptography
- Blockchain
- IoT Security
- Network Security
- Mobile Security
- Data and App Security
- Threat Management
- Standardization
- Privacy
- Software Security
- Hardware Security

For a list of other books in this series, visit www.riverpublishers.com

Visual Communication for Cybersecurity

Beyond awareness to advocacy

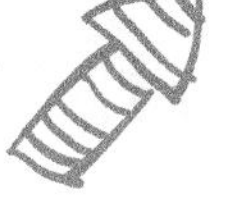

N.E. van Deursen

Manifesto

Present yourself **as you are**. The way you present your norms and values influence how others see cybersecurity.

Talk the language of your audience. Get to know your audience and **connect** with them. Agree on definitions.

Stop bullying your colleagues with dull awareness campaigns. Design policy & products together with end-users. Collaborate with other cybersecurity professionals and speak the same **language**.

Tell stories. Be human. Think human.

Aim higher than awareness. Aim for **innovation** and **advocacy**.

Don't design for security, design for **engaging** security. Visualize your message. Create and share your doodles, sketches, and prototypes together with your audience.

Learn, improve, and **enjoy**.

Cybersecurity is cool. You are cool.

Show it to the world!

Published, sold and distributed by:
River Publishers
Alsbjergvej 10
9260 Gistrup - Denmark
www.riverpublishers.com

ISBN: 9788770220903
e-ISBN: 9788770220897

Title: *Visual Communication for Cybersecurity*
Sub-title: Beyond awareness to advocacy

Introduction

Cybersecurity is at the top of the agenda in business and politics. It is an exciting field to work in with many job opportunities for people from different backgrounds and education. Unfortunately, communication problems are holding back advances in the field. It is difficult to fulfil the many job vacancies because the general public does not really understand what cybersecurity is and what a career has to offer. Professionals working in cybersecurity experience that lay people understand little of security risks and controls. Over the years, security professionals have tried different strategies to promote their work and to improve the knowledge of their audience. For that purpose, communication strategies include avalanches of jargon thrown over lay people or hours of mandatory e-learning modules that will only tick the boxes for the compliance auditor. However, the much-aspired security awareness is only the first step in the process to change behaviour. When we promote a security solution, we need to bear in mind lessons from marketing theory: making your audience aware of it is just the first stage in the marketing funnel to get the audience's buy-in.

Awareness is not the goal.

Many professionals agree that working towards security awareness is a good start and better than doing nothing at all. Even better is to actually change the behaviour of people. However, changing end-user behaviour is not the ultimate goal either. No. We should aim even higher: aim at getting people to look after their peers as advocates of cybersecure behaviour; to have them demonstrating correct security manners and to explain it to their friends, family, and colleagues. Example behaviour within peer groups is far more effective than corporate policy and awareness campaigns.

We need a common language.

Cybersecurity experts disagree amongst themselves about definitions, responsibilities, problems, and solutions. These frictions in the communication amongst experts and awkward communication between experts and their non-expert audience is a complication to the proper protection of valuable information. Cybersecurity professionals need to change the way they present themselves and their services to the public.

Design thinking is the foundation for innovation and engaging security products.

We need creativity and innovation on how to approach cybersecurity, the relationship with lay people, and the interactions with other professionals. We need to design better security products and services together with the people that work with those. When the design processes for security products and risk reduction includes a multi-disciplinary team involving the end-users, the acceptance of the security rules will increase. Design thinking is commonly used in the development of human–computer interfaces, but it can also be applied to other things such as the design of policies, procedures, reports, and training.

With visual communication skills, we can improve collaboration.

The cybersecurity profession needs to be refreshed, with collaboration leading to human-centric design approaches and human forms of communication. Visual communication is useful to explain complex topics and to solve complex problems. It helps to engage audiences. It lights up meetings and it provokes discussion. When done right, visual aids help the audience remember a message and to act on it. When applied strategically, visual communication can contribute to a people-centric approach to security, where employees are encouraged to actively engage in security activities rather than simply complying with the policies.

Cybersecurity education needs to teach visual thinking.

Visual communication has proven its effectiveness in science education, in the communication of health issues, and in legal design. It is gaining popularity amongst cybersecurity professionals as well. Unfortunately, cybersecurity education does not usually include communication theory or design classes. Many experts think that it is not part of their job and is best left to the communication department, or they think that they lack any creative talent.

Reading guide

This book introduces the possibilities of visual communication for cybersecurity. It outlines some of the relevant communication theories and models, gives practical tips, and shows many examples. The book can support students in cybersecurity education and professionals searching for alternatives to bullet-point presentations and textual reports. On top of that, if this book succeeds in inspiring the reader to start creating visuals, it may also give the reader the pleasure of seeing new possibilities and improving their performance.

The book is divided into different parts for readers with different interests. There is no need to read the book from cover to cover; the chapters are organized thematically. If you are interested in academic theory about communication, you will enjoy part II. If you are looking for inspiration and examples of how to use visuals for cybersecurity tasks, go straight to part III. If you want to create something straight away, go to part IV.

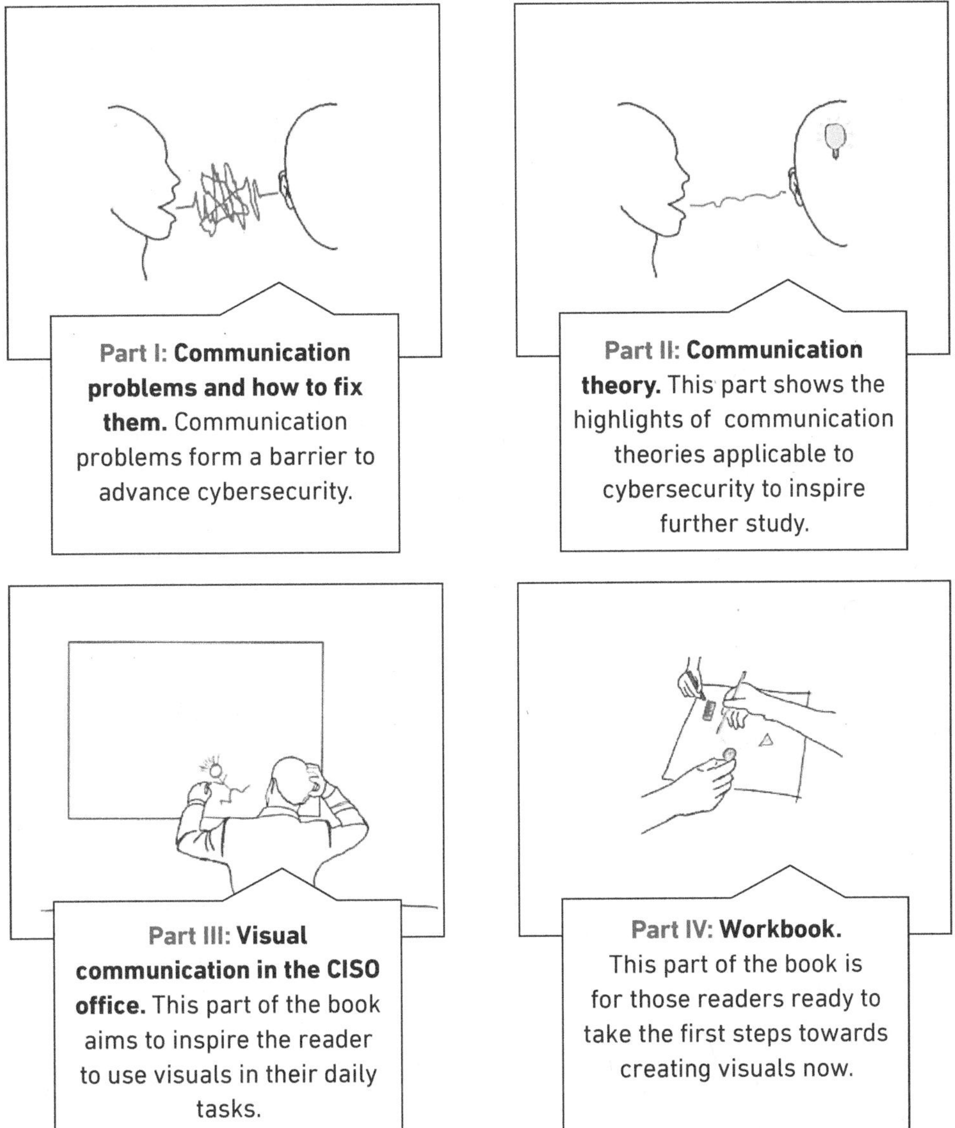

Table of contents

Part I

Communication problems and how to fix them

Cybersecurity communication stress

We need change

Cybersecurity is not only about techies and hackers

Security awareness is nice, secure behaviour is good, advocacy is top

We need more collaboration with different disciplines

No more gobbledygook

Users are not the enemy

Visual literacy & design thinking help:

Security & Privacy by Design

Focus on engaging security for each product, policy, or service

Communication with non-experts

Empathy for the non-expert leads to collaboration

Collaboration

Visuals support collaboration amongst cyber security experts and between different groups of specialists

Fun

Radiate enthusiasm for cyber security

We already have the skills:

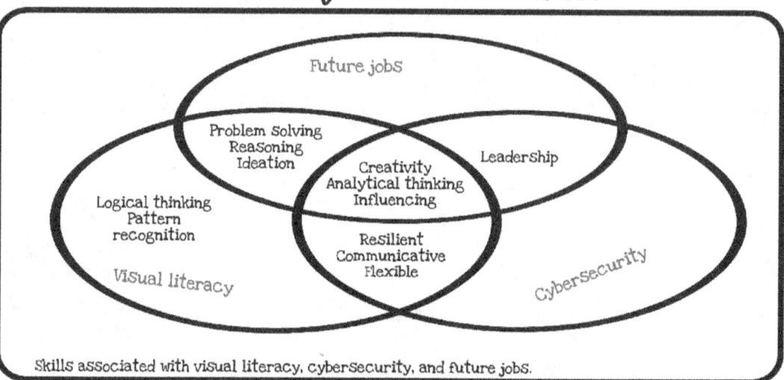

Skills associated with visual literacy, cybersecurity, and future jobs.

We just need to train

1. Image and stereotypes

Should you search online for cybersecurity, images appear of men in hoodies looking at a computer screen. The news stories and videos that you will find speak mainly about cybercrime and warfare. This leads to the public perception that cybersecurity is only about hacking, war, and men wearing hoodies. The image is a disadvantage to the engagement with a broader public and the attraction of a more diverse workforce. The public image suggests that cybersecurity is predominantly male and technical and it does not represent the other disciplines that are working together to resolve complex cybersecurity challenges. This is not only a problem to get the public interested and involved in the risks of their online behaviour but also in filling the millions of open cybersecurity positions worldwide. Some predictions state that this number might go up to 3.5 million by 2021[1]. Apart from the public image, the field also suffers from the image within organisations. The cybersecurity department is seen as the department that says 'no' to everything the business needs and is always checking and spying on colleagues. Cybersecurity colleagues are perceived as 'nosy' and hold up projects with their checklists and requirements. Finally, classic military metaphors that are sometimes used in publications to convince decision-makers 'trigger very specific images in our heads' [2] leading to wrong beliefs.

Cybersecurity protects much more than computers and it involves much more than only computer science.

Cybersecurity is a multi-disciplinary team effort. It is an entire ecosystem not only protecting the confidentiality, integrity, and availability of data, but it also aims to protect our democratic values and even human rights. Knowledge of psychology, behaviour, politics, legislation, and management is just as important as coding, threat hunting, and encryption. Required skills vary from academic to vocational skills in mathematics, computing, law, engineering, social sciences, and politics to protect our economy and our society. Moving away from the image of the hoody to more diverse stereotypes will make it easier to attract a more diverse workforce. Many tasks can be learned on the job and self-starters with skills such as analytical thinking, creativity, communication skills, and complex problem solving can have great careers in cybersecurity.

Cybersecurity professionals need to change the way they present themselves.

The public stereotype image is partly influenced by movies and news stories. Cybersecurity experts can use the same media to influence the public stereotype towards a more realistic image. Cybersecurity is not 'scary, confusing, and dull'[3]. Cybersecurity experts are not there to say 'no' to new business initiatives. To overcome the misconceptions about cybersecurity, we need to establish trust with our audience and employ engaging communication techniques. Cybersecurity is the field that makes new opportunities happen.

Figure 1 An internet search on 'cybersecurity' will show you pictures similar to this theme: a person in a hoody or wearing a mask with one or more screens in a dark, gloomy setting.

2. Awareness campaigns: disconnected and astray

Security awareness is a compliance requirement that organizations need to meet for information security related standards and regulations such as ISO 27001, CobiT, PCI DSS, GDPR, HIPAA, and many more[4]. The majority of these standards and regulations explicitly require organizations to implement a communication and awareness programme for employees and contractors on how to protect valuable information. It has led to a lucrative market for businesses selling awareness training, posters, educative games, campaigns, and so on.

The Oxford dictionary[5] defines awareness as:

1. Knowledge or perception of a situation or fact.
'we need to raise public awareness of the issue'
'there is a lack of awareness of the risks'
2. Concern about and well-informed interest in a particular situation or development.
'a growing environmental awareness'
'his political awareness developed'

Following the Oxford definition, security awareness is a synonym to knowledge and consciousness. There exists no significant relationship between knowledge and secure behaviour[6–8]. Still, many cybersecurity professionals state that humans are the weakest link in security and that they should be trained to change their behaviour. Organizations that plan regular awareness activities such as e-learning, fake-phishing emails, or posters follow the requirements of the standards and regulations, and with that they satisfy the compliance auditors with evidence of activities to 'improve knowledge and concern'. However, if awareness programmes are treated as tick-box exercises, they do not always lead to the desired behaviour[8] and neither does 'scaring, tricking, and bullying users into secure behaviors'[9].

Knowledge of security risks and solutions do not always lead to people actually using those solutions.

When communication is approached as a compliance or policing exercise, the employees in the organization will continue to see cybersecurity as a necessary evil. Auditors are not the only audience that cybersecurity needs to satisfy. The real judges of our work are the people that work with valuable information, such as salespeople, information technology (IT) staff, managers, or human resources staff. Unfortunately, this audience does not seem to like cybersecurity professionals. A survey amongst IT security leaders at 200 organizations in the UK and Germany showed that 84% felt negativity towards their cybersecurity teams from within their organization[10]. One-third even felt that they are being viewed as 'doom mongers'. This negative attitude holds back collaboration and cybersecurity innovation.

To create a more secure workplace, people do not only need to know and be concerned about security, they have to start acting according to that consciousness. In marketing models, awareness is just a first step in the process that leads to the point where customers actually perform an action (buying the product). In parallel to cybersecurity, awareness is only the first step in the process to actually change the behaviour. When we have a security solution, awareness is only the stage with advertisements and campaigns to attract attention. In this context, security awareness can never be the goal. It is only a first step in a longer journey.

Awareness is not the end goal, it is only the beginning.

We need to aim higher. Modern marketing models are no longer satisfied with customers that have bought a product. The stage after that is where customers act as brand advocates and recommend the product to their friends. When we apply this to cybersecurity, we should not treat our co-workers as the weakest link. These people are potential advocates of security. Friends and social processes influence security behaviour more than any other strategy. People change through triggers such as conversations with peer about new threats and by observing how others use a security feature[11–14]. People need to be able to make informed choices in the way they use their computers and share data, to help each other with clear instructions on how to act secure, to make security a routine in their cyber activities, and to have them looking after their peers as advocates of cybersecure behaviour. This starts with trust in cybersecurity professionals and by involving end-users in the design of secure solutions.

WORKMATE BINGO

Security is everybody's responsibility. Be vigilant and be aware that all of the statements in the grid are a security lapse. Be on the lookout to spot if any of your colleagues perpetrate one of these lapses.

Remember you can only stamp your Workmate Bingo card if you point out the security lapse to the person involved or their line manager.

KEEP THIS CARD WHERE EVERYONE CAN SEE IT

GOT A FULL HOUSE? TELL YOUR LINE MANAGER AND CLAIM A PRIZE!

TOGETHER, WE'VE GOT IT COVERED

CPNI Centre for the Protection of National Infrastructure

Centre for the protection of national infrastructure, UK

Workmate bingo[15] Picture copyright: Open Government Licence.

The bingo card invites people to be on the lookout to spot if any of the colleagues blunder with security behaviour. This approach may not suit all cultures. Always use a test audience to evaluate awareness material before you use it. Listen to their response and make changes where necessary to reach the optimum effect of your campaign.

3. Gobbledygook

Cybersecurity is multi-disciplinary but does include, for a large part, highly technical expertise. This expertise is characterized by a lot of jargon. It is often forgotten that most of our audience speaks a different language. Some cybersecurity experts are masters in mysterious language.

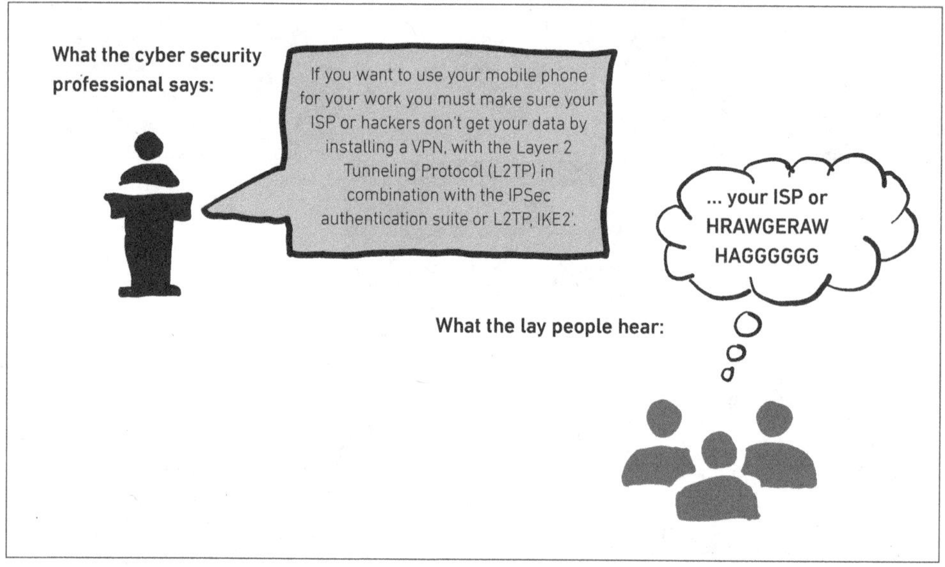

Figure 2 What the professional says, and the lay people hear.

He or she who talks like this to novices is a threat to cybersecurity. On top of this, it is likely that experts will disagree amongst themselves about the technical correctness of statements such as the above. Friction in the communication amongst experts and awkward communication between experts and their non-expert audience is a complication to the proper protection of valuable information. Highly technical concepts should be communicated in a language that lay people understand.

One of the most distressing language problems is that professionals do not agree amongst themselves on terminology and definitions. For non-expert audiences, cybersecurity is a complex and mysterious field that needs clarification. To meet that purpose, various definitions and taxonomies of technical cybersecurity jargon are used by different organizations. This has led to multiple dictionaries[16–19], style guides[20,21], and at least 17 different taxonomies for incident detection and prevention in Europe alone[22]. If even cybersecurity professionals do not use a unified language amongst each other, how will they ever be able to collaborate amongst each other effectively, let alone with people outside their group?

Figure 3 Stop throwing jargon and abbreviations to your audience!

> *How will cybersecurity professionals ever be able to collaborate if they cannot consent on terminology?*

Cybersecurity professionals do not even agree on what their field is called. Is it information security or cybersecurity? The words are used interchangeably by experts and the general public alike. When I am at a party and say that I work in information security, people tend to nod politely and look slightly bored. When I say that I work in cybersecurity, they suddenly pretend to know exactly what I am talking about and start telling stories about hacking and fraud. Cybersecurity as a term seems to connect better with the general public. According to Google Trends, cybersecurity is nowadays more popular as a search term than information security. The many debates about the definition of cybersecurity seem to have in common that cybersecurity aims to protect more than just information. It aims to protect cyber activities and other things that are vulnerable through IT, such as people, cars, machines, or traffic lights. The other things could also be conceptual, such as our economy and society at large, because disruptions in IT could potentially lead to disasters (loss of critical infrastructures), international distrust (digital espionage and propaganda), or mass outrage. Cybersecurity includes the security of IT and information (with the exception of analog information). However, no concise, broadly accepted definition of cybersecurity exists. On top of that, the meaning of the word security is also much contested[23]. This Babylonian confusion of languages evokes discussion amongst experts and this discussion influences the image of the cybersecurity profession. Without a shared language, it is not easy to achieve collaboration and success [24].

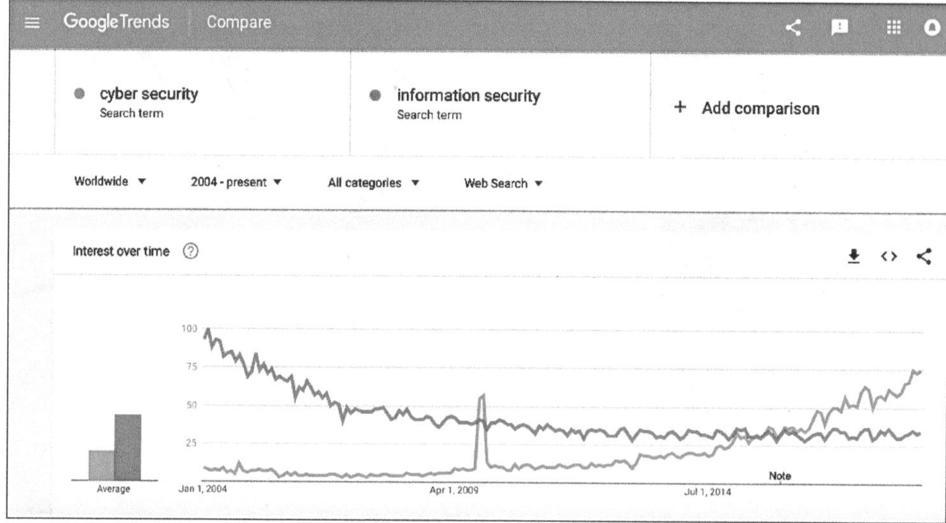

Figure 4 Comparison of the search interest of cyber security and information security. Data source: Google Trends (https://www.google.com/trends).

The complexity of the collaboration between experts from different disciplines is described in more detail in my doctoral thesis[25]. The broadest perspective on our field of work can be described in a multi-layer model. The many risks that we are trying to manage affect more than just ICT and businesses. The global entanglement of people, ICTs, and organisations with cultural norms and ethics calls for risk approaches that are wider than technology or business risks. A cybersecurity risk is not 'a thing' that can be singled out and contained. It is partly related to perception of dangers and annoyances and norms and values. The different perspectives and disciplines that form our work territory are brought together in figure 5. The box in the centre of the figure illustrates the traditional focus on ICT. It is mainly the domain of computer science, mathematics, and engineering. Nowadays, we might want to place Operational Technology (OT) into this layer as well. In this view, security risks are technology risks that threaten the confidentiality, integrity, or availability of systems. These risks can be contained through the installation of artefacts and mechanisms and through the use of standards and checklists. Around the centre, the growing insight from business, sociology, politics and humanities complement the technical scope. These different streams strengthen and inspire the field. Security thinking has expanded from the concept of technical containment to socio-technical complexity. This does not mean that one could replace another: the different perceptions are complementary to each other and all of them are needed to understand risks. Risks include the whole of technology risks, business risks and society risks. A technology risk (such as the risks that a critical infrastructure stops working because of a technical issue) could cause a business risk (when organisations relying on the infrastructure cannot complete their production). In turn these risks could cause risks to people and groups in society. To illustrate the disappearing boundaries around environments and systems, the box around the technical area has dashed lines and the text on the outside is not framed at all, as a symbol of the pervasiveness of security issues.

The outside area of the figure is related to risks in society. These risks can be identified and better understood by studying information society discourse. From studying the works of information society thinkers such as Ulrich Beck, Manuel Castells, Jan van Dijk, or David Lyon, we can learn that the scope of security risk is global and infinite, through the connections in socio-technical networks. Furthermore, the risks can be economic (influencing stock markets), personal (identity theft, social inequalities, damaged social relations), political (bringing down governments, or damaging international relations), and have the potential to influence basic human rights (the right to be left alone, freedom of speech). Information society theory teaches us that the perception of societal security risks is likely to be socially constructed by international power and power struggles, (lack of) public information policy, mass media and culture.

We face a difficult, maybe even impossible, task in bringing all these disciplines together in a field that we are trying to scope with the term cybersecurity. Yet, as technology risks could pose a threat to our human rights and physical safety, we'll have to find a way to work together across disciplines and nations.

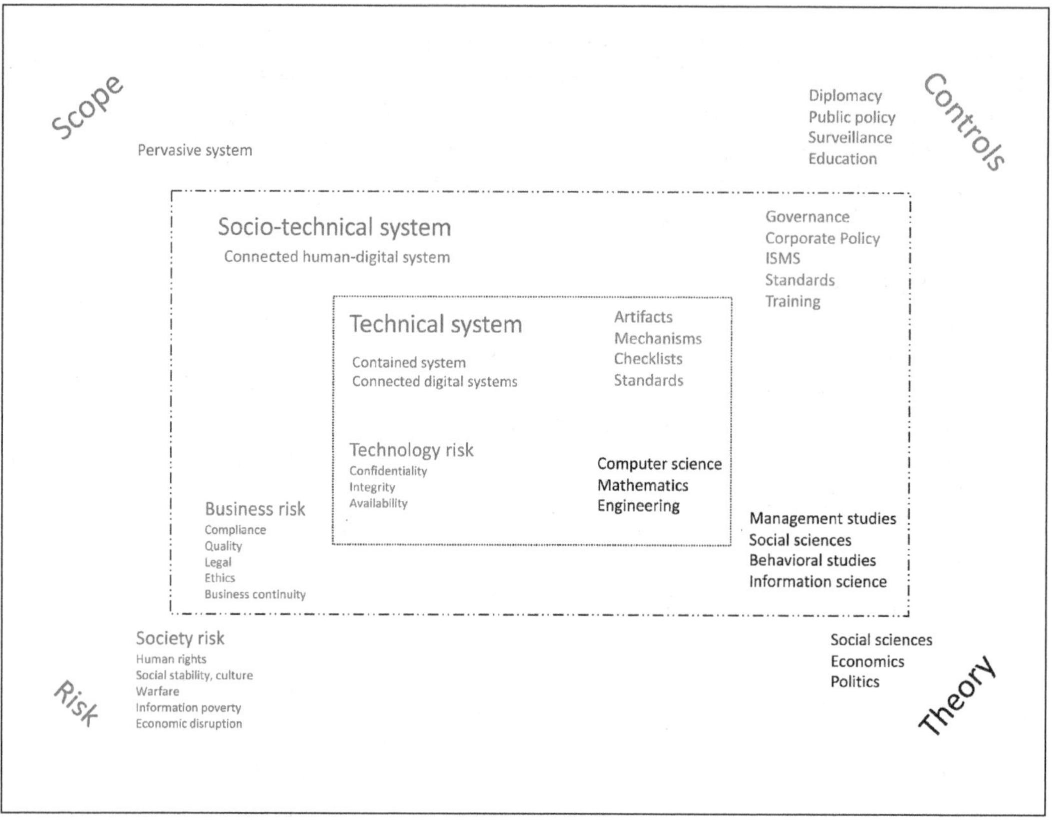

Figure 5 Scoping the cybersecurity field.
Adapted from the author's doctoral thesis[25]

4. The way forward

The aforementioned communication problems have resulted in a large disconnect between cybersecurity professionals and their stakeholders. Cybersecurity is *an* exciting field but has a hard time shaking off a misplaced reputation. Visual communication is one of many other tools that can be used to improve this disconnect.

Some may think that visual communication only means drawing pictures. They may say that drawing is frivolous and unprofessional and that it is something that only 'creative types' do. However, visual communication is not about drawing pictures, it is about applying meaningful visual tools. Those tools can also be stories, as stories help the audience to visualize a mental picture. Other tools are the use of colour, typography, and text size option to visually structure information. It helps us in many ways:

- It orders our thoughts.
- It helps to recall information.
- It increases insight in patterns.
- It helps to see the big picture.
- It supports decisions.
- It communicates information.
- It generates new ideas and inspires.
- It makes a subject approachable.
- It provides for a relaxing means to find focus.

Visual communication is one of the components that form visual literacy and is inherent to design thinking.

4.1 Visual literacy

Visual literacy is the ability to read, comprehend, create, and use, and the insight to plan visual language. It is grounded concepts such as visual perception, visual language, visual learning, visual thinking, and visual communication[26]. It is a topic of study and discussion in different fields amongst which are librarians, designers, computer experts, educators, and biologists. In recent years, with the growing use of social media and multi-sensory devices, visual communication is finding a place in education and businesses. It is an interdisciplinary and multi-dimensional area of knowledge and skills that involve cognitive functions such as critical viewing and thinking, visualizing, deduction, reasoning, and construction of meaning. Unfortunately, it is hardly included in the curriculum of cybersecurity education. This is a big gap in the development of the profession since its inclusion develops skills such as problem solving and collaboration. Cybersecurity professionals need to learn how to use engaging communication techniques, such as showing enthusiasm and making security relatable[3].

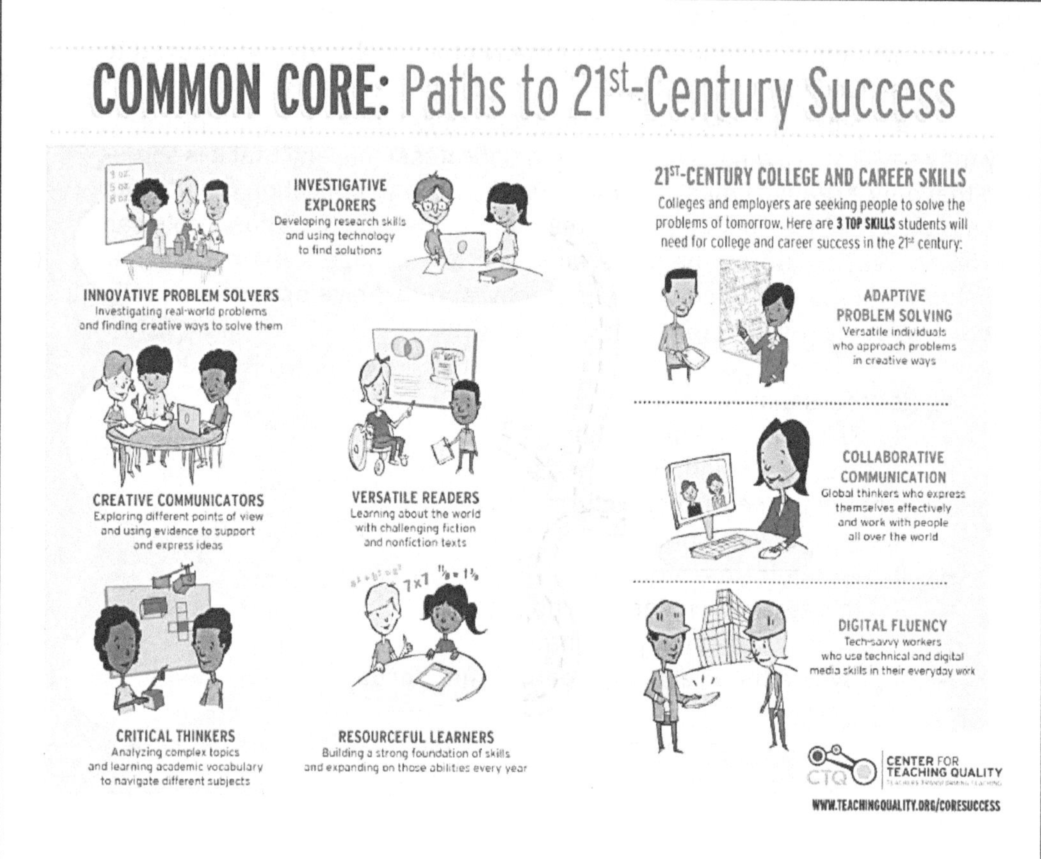

Centre for teaching quality

Common Core: Paths to 21st-century success[27] Copyright: permission to print from the Centre for Teaching Quality

This poster is an aid for teachers to explain why visual skills are essential to college and career success in the 21st century. The arguments apply also to cybersecurity education and career success.

Some people think that it is not part of their job or that they lack the creative talent to take this path. However, communication is part of every cybersecurity professional's work, and with some training, everybody can sketch visual aids. It is a matter of building the confidence to try something new, something different. Visual communication is especially appreciated amongst millennials[28], although even digital natives need training to create meaning from visuals[29].

Visual literacy needs training.

As children, most of us drew simple pictures of stick men and square houses. As adults, many of us stopped drawing our ideas and find it difficult to draw. With some practice, it should be possible to overcome the reluctance to discover the impact of visual communication based on simple drawing skills. Visual literacy skills are measurable and capable of development and improvement[30]. Training visual skills gives a lot in return. Visual communication gives people the following[31]:

1. Power (a form of advanced thinking that gives us better information comprehension, retention, and recall).
2. Performance (a tool for deeper group engagement, problem solving, meeting efficiency, and shared memory).
3. Pleasure (a personal gift to focus and relax).

Visual communication skills vary from putting a doodle on a napkin and stick figures on a white board, to artistic speeches, professionally designed graphics, and virtual reality. Some businesses take it very seriously: Google employs an official Chief Doodler, Microsoft has a Chief Storyteller, and at Ford, there is a Master Clay Modeler who builds clay models for each new car. These types of jobs demonstrate that there is another dimension beyond the well-known bullet-point presentations.

Pictures are proper business attire.

The World Economic Forum Future of Jobs Report[32] states that creativity as a skill will increase in value. The list of top skills further includes complex problem solving and critical thinking. Visual communication is a tool to support exactly those skills. Furthermore, there is an overlap between the top skills of future jobs, the most required cybersecurity skills[33], and skills attributed to visual literacy.

Learning styles are often divided into four types: visual, auditory, read/write, and kinaesthetic. The visual learning style, often referred to as the spatial learning style, is a way of learning in which information is associated with images. Research into preferred learning styles keeps finding that most people

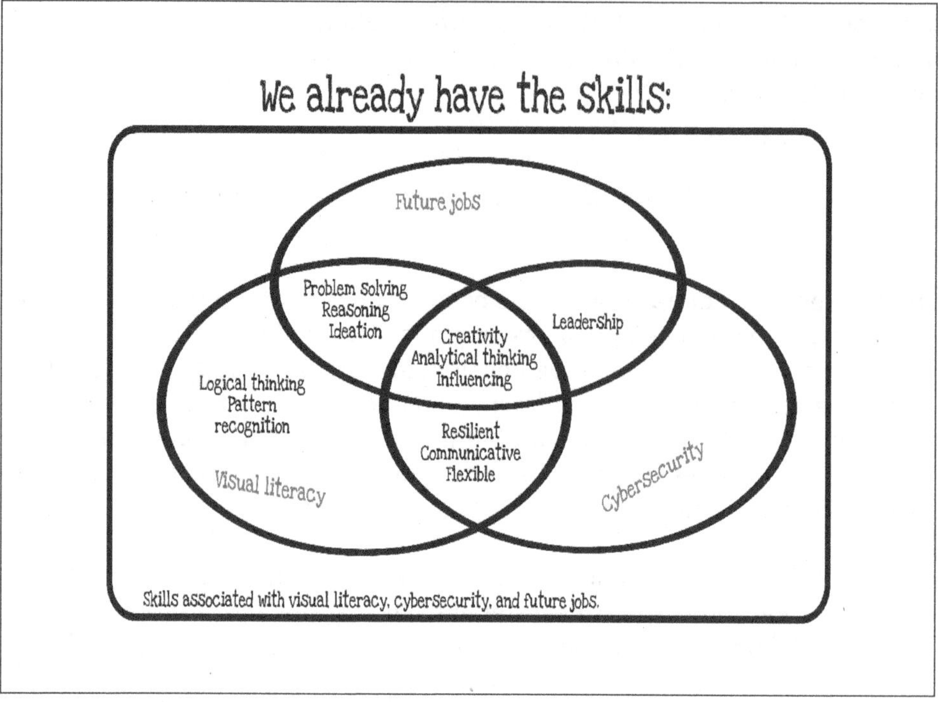

Figure 6 Future job skills.

prefer combined learning styles[34]. It demonstrates that visuals are a much-appreciated addition to communication style and should not be forgotten.

The proverb 'a picture is worth a thousand words' is often used to promote visual thinking. It leads some to think they need to replace all content with pictures. That is not necessary and often not helpful at all. Most of our information consumption is text-based and we need words to communicate. A picture cannot always replace all words, but it can help to reduce the number of unnecessary words and an attractive layout can draw attention to something important. An image might support and strengthen a text and make it more interesting. Some images (such as journalistic photos) can also provoke words and spark discussion about an event. Other pictures help to convince an audience of a version of the truth that is written down in lengthy documents (such as in legal files or in fraud investigations). Text and images complement each other.

In the case of data privacy and security, even the GDPR suggests the use of visual aids such as icons. Article 12 of the GDPR requires the provision of intelligible and easily accessible information on data practices. Article 12(7) even suggests the use of icons 'in order to give in an easily visible, intelligible and clearly legible manner a meaningful overview of the intended processing'.

4.2 Design thinking

Design thinking is an iterative process in which we seek to understand the people for whom we are designing the products or services. The process helps us to develop empathy with the intended users. The engineering behind computers is beyond comprehension to a large proportion of mankind. Fortunately, most people do not need to know all of the technical details. However, they should be engaged with the design of new solutions and be ready to accept changes to existing systems. They also need to be able to reproduce new instructions to others. With some practice, specialists should all be able to communicate complex topics to end-users and to encourage them to tell it on. Most problems and incidents in cybersecurity are caused by human behaviour and not all risks can be controlled through technology. Design thinking is an approach to design security products and services in collaboration with multiple disciplines and end-users from the beginning. Cybersecurity education currently teach little about user centric design although there are rare initiatives to change this[35]. Engineering studies focus on system development, developers learn about coding and business students learn about risk management. Design thinking is most likely to be of interest to cybersecurity professionals with a background in software engineering: design thinking is the foundation of several software development methods. However, design thinking is not only suitable for the development of software or human–computer interfaces but can also be applied to the development of corporate policies, procedures, and training programmes. It also applies to the design of visual aids in communication processes. Design thinking by its nature is a very visual process, supported by visual tools and techniques such as brainstorming, sketching, prototyping, experimenting, and testing. It helps to design usable security.

Design thinking provides the processes to create visual communication and better security solutions.

Security by design means that the development process for security products (these could be hardware, software, as well as the accompanying procedures) aims to make systems as secure as possible from the beginning of the design process and continues to test and improve security controls throughout the design cycles. Design thinking is a process in five stages[36]: empathize, define, ideate, prototype, and test. The stages are not sequential: the process is iterative in nature. Design thinking assumes that you may not reach your final solution in one go but that you take more experimental approaches to prove ideas early on and then adjust based on user feedback. The user takes central stage: they help to seamlessly blend cybersecurity controls into their environment, instead of cybersecurity experts wishing that users would just follow technically perfect security controls that they impose on them.

Design thinkers have a few distinctive qualities such as empathy, optimism, integrative thinking, experimentation, and collaboration. These qualities support the way forward to design better solutions. In this light, designers can be seen as problem solvers. Cybersecurity problems are often wicked: they are ambiguous problems:

- A solution that works today might not work in the future.
- A security control that works in one organisation might not work in another organisation.
- Most cybersecurity solutions require a lot of people to change their mindsets and behaviour.
- Some security problems are symptoms of other problems.
- New threats are often novel and unique.
- There is not enough time.
- The problem is never solved definitively (100% security does not exist).

Wicked problems can't be "fixed". Approaches should be focused on how to best mitigate their immediate impact. Finally, wicked problems require an interdisciplinary approach with an understanding that no quick result will be forthcoming. Addressing wicked problems is time-consuming and iterative, requiring long-term dedication[37]. This is one of the reasons that design thinking as a problem-solving method seems promising for the cybersecurity field.

5. Conclusion

It may seem ambitious to write a book that combines two fields that are fuzzy, widely misunderstood, multi-disciplinary, as well as multi-dimensional. At the same time, cybersecurity and visual literacy have in common that they require similar cognitive functions such as critical thinking, visualizing, deduction, complex problem solving, and reasoning. Cybersecurity professionals are likely to possess the skills required for visual communication but lack the training to confidently use it. Cybersecurity suffers from communication problems such as the absence of a common langue and an obscure image. This is impeding the progress of the field and obstructs the path from security awareness to advocacy.

Recently, a more positive vibe towards cybersecurity amongst the general public is certainly noticeable. Consumers, educators, parents, and politicians ask more questions and debate publicly about cybersecurity and privacy. Mass media and social media have contributed to the dispersion of stories of cybercrime and the interest from a wider audience. News stories about data breaches, identity theft, or cyberattacks by foreign governments attract attention. One explanation is that news stories offer a learning experience and, as a result, the public becomes more informed. The media has a wide reach to different audiences and many different channels to deliver news stories, opinion, and background articles. Communication theorists attribute the influence of mass media on public opinion and behaviour to the power of visual illustrations. Mass media traditionally use visual communication to illustrate, attract attention, and to explain a sequence of events. Cybersecurity professionals in organizations could benefit from these techniques to gain the interest from the audience within their organization. Unfortunately, cybersecurity education does not usually include communication theory or visual design classes. Producing clear, organized visual communication requires visual insight. The problem is that there remains an outdated bias that visual production is a vocational and not an intellectual process[38], and at the same time, there remains an outdated bias that cybersecurity is an intellectual and not a vocational profession. The reality is both are needed to create progress in cybersecurity.

Visual communication is a tool that supports each step of the way that leads us from the first impressions of cybersecurity to knowledge and awareness to advocacy. When applied strategically, visual communication can contribute to a people-centric approach to security, where employees are encouraged to make informed security decisions, take personal accountability, and actively engage in security activities, rather than simply complying with the policies[39]. This will result in cybersecurity products and services that are transparent, accessible, visually clear, useable, understandable, useful, and engaging.

Ultimately, this will advance the development of the discipline.

Part II

Communication theory

Communication Theory

Cybersecurity needs communication theory

- To get to know your audience
- To understand your audience: how they learn and are persuaded
- To plan communication activities and select channels

- To design for attention
- To better liaise with communication specialists
- To create better content

Knowing and understanding your audience

Theory area:	Key takeaways:
How people learn	• Tailor security education to the learning objectives. • Mix visuals, spoken words, written words, and hands-on for maximum effect
Context where communication takes place	• Adjust communication style to match the organisational structure • A shared culture = a shared language • Stakeholders influence success • Trust in employer motivates compliant behaviour• Messages must be relevant and repeated
How people are persuaded	• A visual tool has more effect when the design is coherent • Information should fit with what is already believed • Persuasion needs trust and reassurance • the strongest influencer is the social group that a person belongs to

Basic theory and models

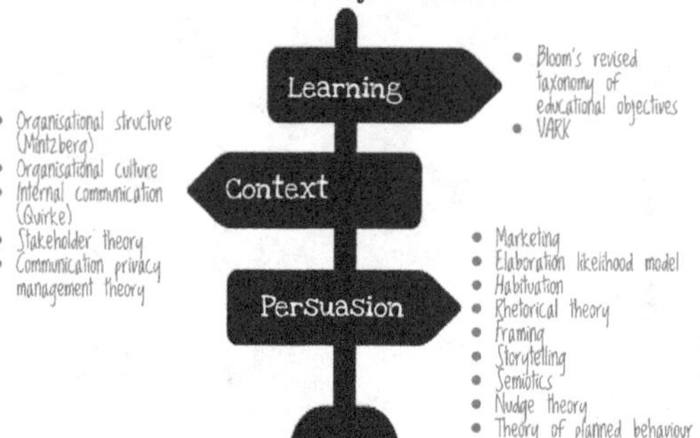

Design for better content and attention

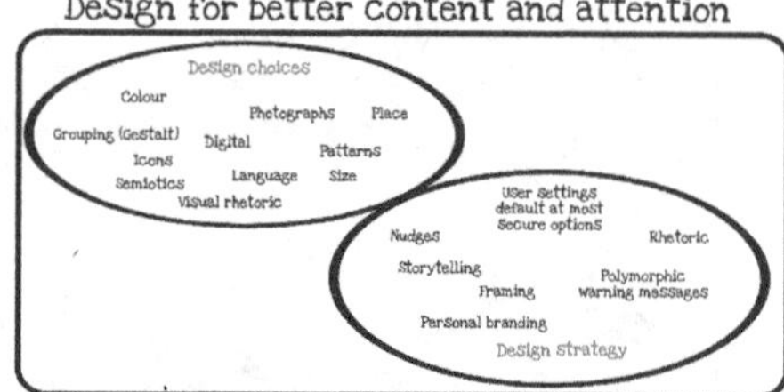

Awareness is at the bottom of these models

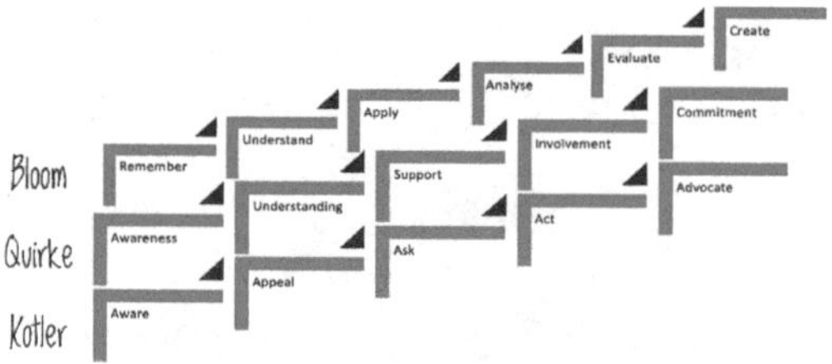

Improved Communication gives us something much better than awareness:

6. Introduction

This part of the book briefly introduces some of the theories and models studied in communication science and how they can be applied to cybersecurity. The content may be a bit heavy for some readers. It is not very visual and frequently refers to academic publications. In some business situations, it helps to have this background information about communication models and theory. Basic knowledge of communication theory will help cybersecurity professionals in several ways:

1. To get to know their audience
2. To design the best way to teach their audience
3. To select communication channels that will reach their audience
4. To plan communication activities
5. To design and create better content
6. To better liaise with communication specialists
7. To understand why awareness is not good enough

The overview on the previous page summarizes this part of the book. The theory and models are grouped into three areas: how people learn, the context in which communication takes place, and how people are persuaded. The overview of theories introduced in this part is far from complete. There are many more models that could apply to cybersecurity communication. Unfortunately, research into applying those models to a cybersecurity context is scarce. There is still much to learn, and hopefully, this book section can be of interest to researchers who are at the beginning of a research project and are looking for an overview of communication theory applied to cybersecurity.

7. How people learn

Security awareness programmes aim to inform and train employees so that they are able to perform their tasks in a secure way. In many organizations, these awareness programmes are created by the security staff, sometimes in collaboration with communication specialists. These awareness programmes often include training programmes and, therefore, cybersecurity professionals will benefit from some knowledge of learning theory.

7.1 Bloom's taxonomy

In the 1950s, Benjamin Bloom, together with a group of scholars, developed a classification of educational goals known as Bloom's Taxonomy of Educational Objectives[40]. In 2001, key elements of the taxonomy were updated to reflect more relevance to 21st-century educational goals[41]. The taxonomy classifies three areas of learning: the cognitive domain (knowledge), the affective domain (emotion), and the psychomotor domain (actions). The cognitive domain consists of six levels of learning objectives:

Table 1 Bloom's levels of learning objectives.

Bloom's level (version 2001)	**Learning objective**	**Related verbs**
Create	Produce something new	Design, forecast, reorganize, construct, conjecture, develop, author, investigate
Evaluate	Justify a stand or decision	Argue, defend, select, support, value, critique, weigh, estimate, classify, assess, check, appraise
Analyse	Connect ideas or break them down.	Organize, relate, inspect, investigate, compare, distinguish, experiment, question, test, examine
Apply	Use information in new situations	Execute, compare, implement, solve, use, demonstrate, interpret, operate, schedule, sketch
Understand	Explain ideas or basic concepts	Classify, describe, discuss, explain, identify, locate, recognize, report, select, translate
Remember	Recall facts and basic concepts	Define, duplicate, list, memorize, label, locate, repeat, state

Table 2 Bloom's levels of learning objectives applied to cybersecurity.

Bloom's level (version 2001)	Security learning examples for professionals (adapted from[44])	Security learning examples for end-users (adapted from[43])
Create	Formulate a security risk assessment and disaster recovery plan	Write a new policy item to prevent users from putting sensitive information on mobile devices
Evaluate	Classify threats and countermeasures based on a risk assessment	Critique these two passwords and explain why you would recommend one over the other in terms of the security it provides
Analyse	Identify and analyse project risk and perform qualitative and quantitative analyses	Which of the following security incidents involving stolen passwords are more likely in our company?
Apply	Assign appropriate physical versus logical and centralized versus decentralized access control in various scenarios	Use the appropriate application to change your password for the financial sub-system
Understand	Explain auditing, asset management, standards, and enforcement when managing networks	Why should non alphanumeric characters be used in a password?
Remember	State the user account password requirements	What is the definition of access control?

Some researchers state that traditional security awareness training can only reach the bottom three levels of learning: remember, understand, and apply[42]. However, to ensure cybersecurity in our modern digitized society, every person needs skills at the highest level possible to handle their digital activities.

Bloom's taxonomy might assist in tailoring the level of education to the needs of the audience with different forms of communication that are not often explored by cybersecurity professionals. 'For example, simply teaching an individual what a password is, would lie on the remember, and possibly understand level(s) of Bloom's taxonomy. However, [...] to understand why their own passwords [...] should also be properly constructed and guarded might lie as high as the evaluate level of the taxonomy'[43(p284)]. This requires people to be able to compare passwords and to explain which one is stronger.

Bloom's taxonomy could also be useful in the design of cybersecurity education for professionals. It helps to identify areas where cybersecurity topics could be added to IT courses.

The other domains from Bloom's taxonomy (affective and psychomotor) are highly overlooked in a cybersecurity context. Skills in the affective domain target, for example, attitude and the psychomotor domain focusses on change in behaviour and skills. To improve the gap in knowledge about cybersecurity learning in these domains, we might find connections in theory of psychology and communication, such as those described in the remainder of this book section.

7.2 Learning styles

Learning style theories claim that individuals learn differently and classify people according to their preferred style of learning. A plethora of such models sprouted from the 1970s onwards. These theories are much contested and criticized for the lack of scientific evidence that individual learning styles even exist. Still, many teachers support these theories. In 2014, 90% of surveyed teachers in five countries (the United Kingdom, the Netherlands, Turkey, Greece, and China) believed that individuals learn better when they receive information tailored to their preferred learning styles[45].

One of the theories divides learning into four modalities that are used for learning information: visual, aural, read/write, and kinaesthetic (VARK)[46,47]. Aural or auditory means a preference for information that is spoken or heard. The read/write modality refers to a preference for written words. Visual modality includes the description of information in maps, diagrams, charts, and so on. It includes designs, patterns, shapes, and other symbols. Remember that not all pictures are visual. Use of a screen (think presentation slides) does not mean that the material on it is visual. The message can be text (bullet points), and just because it is shown on screen does not mean it is a visual message. The fourth modality, kinaesthetic, refers to the preference to experience and practice as a learning style. Learning by doing or watching videos of themselves doing it are most effective for people with this preference. With the help of a questionnaire, people can find out what works best for them. People may switch between modes in different situations or prefer a combination of two or more modes. Different researchers that used the VARK questionnaire found that the majority of the respondents prefer a combined style. This implies that teaching methods should blend the four styles to improve the chance of maximum effect.

VARK and communication have strong links[48]. Interpersonal communication is embedded in the four styles. People inform, persuade, or build relationships the best with their preferred style. Mismatches in VARK profiles between people can lead to conflicts and misunderstanding. Marketeers use the VARK principles in television commercials when they appeal to all modalities. Other forms, such as a video, to explain how a product works appeal more to kinaesthetic learners. The read/write modality is needed to read Internet sources, as many websites are textual.

Cybersecurity professionals can take away from this theory that for effective communication at conferences, workshops, in marketing, or team building situations, the combined use of visuals, text, videos, games, and physical activities may be more effective than relying on one single modality. This is also relevant for the design of awareness training. A study in Australia looked into the question why the staff at a specific bank demonstrated higher levels of information security awareness than the general workforce in other industries[49]. Research into the explanation of this phenomenon found no correlation between frequency of training and the level of awareness. Therefore, the explanation was sought in the type of training that the staff received and the way that it was designed. The team that developed the training regime consisted of behavioural experts and educators. Being aware of different learning styles, they used different channels and media to communicate the topics of interest, including video, posters and flyers, and informative emails. Furthermore, all employees must pass a mandatory training course before gaining access to the ICT systems. Management should focus less on the frequency of awareness training, and more on the design of the types of training.

7.3 Security education, training, and awareness

Danish researchers[50] recently looked into the body of academic literature about security training and awareness programmes. Where in the early days of information security the focus of such programmes lay on technology, since the beginning of this century there is more attention to the human factor. Unfortunately, research after research finds that even with the human in mind, many awareness programmes keep failing to achieve the desired results. Therefore, the researchers evaluated 42 research papers about information security and human behaviour and distilled factors that affect the success of security education, training, and awareness programmes. These factors could serve as input to design future programmes.

The researchers divide the factors into four dimensions:

- Knowledge: influenced by factors such as: content, use of respectful language, and the experience of the distribution of knowledge.
- Attitude: influenced by factors such as: beliefs, personal norms, and knowledge.

- Intention: influenced by factors such as: organisational control, perceived threat severity, rationalization of actions, perception, and motivation.
- Behaviour: influenced by factors such as: environment, cognitive processes, and perception of environmental influences.

The dimensions are related to each other. For instance, knowledge is an important predecessor for intention and for attitude. Knowledge in it turn follows from experience, sharing, and collaboration. In their paper the researchers list many more direct and indirect factors on each dimension than the ones outlined above. The combination of knowledge, attitude, and intention influence people in their behaviour and their decisions to comply or not to comply with security policies. Organisations are likely to achieve more success with their awareness and training programmes if they take these factors into consideration.

7.4 Lessons from phishing simulations

The popularity of simulated phishing campaigns as part of corporate cybersecurity awareness training is still rising. A phishing simulation is when an organization sends a mock phishing email to employees. It's designed to provide feedback if employees click the link included in the phishing email. Actions such as opening the email or clicking on the link are analysed to evaluate how susceptible the people are to email phishing campaigns and to guide further awareness training. The theory behind this approach is that staff will learn from their mistakes and that the next time they receive a suspicious email they will be more vigilant. Some academics call this bullying[9] and there is even an example of a trade union publicly disapproving of misleading workers in a financial institution[51].

Research into the effectiveness of this kind of campaigns delivers mixed results. A study amongst 1,350 university students in the USA even found an unexpected result: they observed greater user susceptibility with greater phishing knowledge and awareness[52]. Another study[53] amongst nearly 11,000 students and university staff saw a drop of the number of people that clicked on phishing emails in the second round of an awareness campaign. However, the numbers were still high, considering it was the second round of a repetitive campaign combined with an announcement that there was a phishing scam.

A recent study in Thailand[54] used a mixed training format approach to educate participants in the phishing simulation. Before the training the participants received several simulated phishing emails. Then, the researchers used combinations of video-based training, game-based training, text-based training, and instructor-led classroom to educate the participants. After the training there was another round of simulated phishing emails and there was a significant decrease in click-rates. However, the second round was done immediately after the training, so there is no real proof of long-term effects. Furthermore, there was no difference in effect after specific combinations of training methods, so none of the training method combinations performed better than another. The participants had different opinions on the training methods, therefore awareness campaign designers should include a variety of learning methods to meet the learning style preferences of the participants.

What studies such as those mentioned above lack is the measurement of effectiveness of the training over a longer period of time. Several studies suggest that the effect wears off after 28 days[55], but not enough studies that include repeated campaigns over a longer period of time have been conducted. One study in healthcare[56] measured the effect of 20 simulated campaigns, combined with mandatory phishing training, over nearly three years. They found that there was some reduction in click rates, but the number of susceptible employees remained high. An additional mandatory phishing training was introduced for a specific group within the participants after 15 rounds of phishing campaigns. The training did not lead to a significant decrease of the click rates in round 16 to 20 by this group and there was no difference in results from the group that did not undergo the training course.

We need a lot more evidence before we can draw any conclusions about the effectives of phishing training through mock-phishing simulations and additional training. We are only just beginning to learn about teaching cybersecurity concepts and the long-term effectiveness of training and more research into the effect of training should be conducted.

8. Context in which communication takes place

Effective communication requires knowledge of the environment where the communication takes place. Most cybersecurity professionals will focus on communication within their organization. Internal communication is a continuous process of exchanging and interpreting messages between members within that organization. This type of communication strongly depends on organizational structure and culture. Before planning communication activities, it is important to understand the organization.

Organizations can be described by six variables[57]:

1. Goals: why does the organization exist?
2. Strategy: how the organization reaches its goals
3. Structure: the division of tasks
4. Culture: unwritten rules that exist within the organization
5. Technology: machines, ICT, procedures, and tasks
6. People: people that work in the organization

Although all six variables influence internal communication, the following paragraphs are limited to structure and culture because these topics are highly relevant to empathize with the people that we want to engage with.

8.1 Organizational structure

The business goals and the division of labour in an organization influence security. The organizational structure can be an obstacle for security as well as a source for change. Take, for example, the concept of security by design, expected to be a common practice in software development. However, the reality in organizations does hardly match the best practices for secure software development.

A study in Canada explains the reason for software development teams not following the best practice for a secure software development. Most organizations follow their own software development method, and the best practices for secure development conflict with their team members' roles and responsibilities. Many organizations find it unreasonable to change their structure in order to comply with these best practices[58]. Among the teams that were observed, only one considered security as a part of the design stage of the development process. Many of the other teams stated that they are not responsible for security and they are not required to secure their applications. In fact, some developers reported that their companies do not expect them to have any software security knowledge. The study also found that these organizations usually appoint a specific person or team to be solely responsible for security. This leads to ad hoc processes and violates the best practice that security is a shared responsibility.

Another organizational obstacle to secure software development was found by German researchers. They studied a group of software developers during and after their encounter with security consultants to find out how their organizational routines changed so as to integrate security work[14]. They realized that consultancy had effects in the short term, but sustainable change like the adoption of new tools or practices did not follow. The researchers related the main cause to agile approaches: the development team prioritizes explicit requirements to deliver what is required in the contract. And as security is always an implicit expectation and not an explicit business goal stated by the management, it did not receive lasting attention. The researchers conclude that the success of integrating security into organizational routines depends on the specific organizational setting. The way business management perceive security influences how projects prioritize security in the organization.

The structure of the organization is an important factor for cybersecurity communication. Mintzberg[59] described five main parts of an organizational structure, and the most dominant of these parts defines what type of organization it is. The basic five parts are:

1. At the strategic top, we find top management and its support staff.
2. At the bottom of the organizations is the operative core, the workers who carry out the organization's tasks.
3. In the middle, we find the middle- and lower-level management.
4. The cluster of functions that is formed by engineers, accountants, planners, ICT, and researchers is called the technostructure.
5. The support staff are the people who provide indirect services such as maintenance, legal council, and food service.

According to Mintzberg, there are seven organization types, depending on which of the five parts plays the dominant role. Mintzberg's basic organizational configurations are:

1. The entrepreneurial organization: it has a flat structure and a large strategic top.
2. The machine bureaucracy: a heavy middle management and standardization of work processes.
3. The diversified organization: the organization is divided into divisions which focus on their own products.

4. The professional organization: the employees are independent, but there are many guidelines and procedures.
5. The innovative organization: the structure is adaptable and project based.
6. The missionary organization: the company culture is based on ideology.
7. the political organization: a power-oriented organization. There is no formal structure and it is unclear who is responsible for what.

When the cybersecurity team plans communication activities, it helps to recognize what the shape of the organization is and where the intended audience is placed. For example, internal communication in a machine organization is often standardized and formal. It is a type of organization that maintains a lot of manuals. Communication is predominantly written and formal. Even bottom-up communication is formal and meetings often result in rules and procedures. Communication is different in an entrepreneurial structure. Here it is informal, interactive, and directly between the boss and the employee. The power lies on the operational level in a professional organization and employees are engaged in the creation of policies and procedures. Most staff are very independent and hardly communicate about their tasks.

In cybersecurity, we sometimes see an approach to communication that does not fit the structure. This happens, for example, around the implementation of management systems for information security (ISO 27001). Security officers often formalize the communication as if they are in a machine organization, even when, in reality, they find themselves in an innovative or entrepreneurial structure. Or they impose policy and procedures on a professional organization without consulting the independently working staff. Large organizations may even have different structures across different divisions. Knowing these structures is beneficial for planning the communication about cybersecurity and for adjusting the style and activities to what suits the different target groups.

8.2 Organizational culture and policies

Organizational culture is the dynamic process that gives meaning to our workplace. It is expressed through informal communication, corporate politics and influencers, irrational decisions, feelings, emotions, and unwritten rules. Organizational culture is formed through the ideology of the organization (values, ideas, and preferences) and manners (behaviour, rituals, stories, architecture, branding, and logo). It is not a static situation but a continuous construction by people. It can be changed, influenced, and even designed by the management[60].

Many consider the organizational culture as a main factor for cybersecurity risks coming from inside the organization (insider threat). The European Union Agency for Network and Information Security (ENISA) defines cybersecurity culture as: 'the knowledge, beliefs, perceptions, attitudes, assumptions, norms and values of people regarding cybersecurity and how they manifest themselves in people's behaviour with information technologies'[61(p40)]. In the past, researchers focused on how strengthening regulations[62] and reporting abnormal behaviour[60] influence security culture. In this way of thinking, it was believed that with policy, compliance frameworks, assessments, and punishments[63], a security culture will develop itself. This vision is still seen today in best practises such as ISO 27001 and in recommendations from organizations such as ENISA although there is evidence that this ideological approach is unlikely to be effective in some types of organizations[64] and that sanctions work counterproductively[65]. On top of that, cultural risks will vary across an organization, and it is not likely that one homogeneous security culture can be established at all[66].

Figure 7 Word cloud of the text in the ISO 27001:2005. The larger words are used the most

The culture of an organization is an important variable for successful cybersecurity improvement. From a communication perspective, changing a culture requires one or more of the following communication actions[57]:

1. One-on-one conversations about what needs to be changed
2. Demonstrate successes with the new way of work
3. Investigating incidents together and discuss options
4. Make unwritten rules known
5. Pointing out exemplary behaviour and role models
6. Rewarding desired behaviour

These actions may help to convince people to change over time. People usually do not want to be forced to change, but they might want to try if they are convinced of the advantages. Participation of local group representatives in corporate policy development and cultural design programmes will contribute to the success of those programmes. Investigating incidents together with end-users contributes to the understanding of the consequences of actions and end-user behaviour. On top of that, this participation will help to achieve what would be one of the biggest changes in security culture: the evolution towards a common language. Where people collaborate for a long time, they tend to develop a shared language that helps them to communicate more effectively. When people from different disciplines work together, it takes time to understand each other. When such a team actively creates a language together, it helps the team members to feel more constructive and it gives focus to all participants. Language reveals the relationships between individuals, groups, organizational success, and even individuals' career prospects[67]. To understand the culture and sub-cultures in an organization, it is important to understand the language through which people communicate with colleagues. In this area lays an opportunity for cybersecurity professionals to create a universal language, at least for their own organization, and share it through dictionaries and style guides. Eventually, this will improve the relationship between cybersecurity professionals and their co-workers within an organization, as well as the relationship with stakeholders from outside the organization.

'Shared commitment only comes after shared understanding. Through the process of developing a shared language between individuals, their working (and personal) relationship is often enhanced'[24].

The corporate culture and how it translates into security policies affects the level of commitment from the employees to comply with these policies. One suggested cause of problems with policy is the style and wording of the information security policies. Some policies can be as long over a hundred pages and in a technical writing style and they are likely to be ignored[68]. A shorter policy gets across the message about the need for information security to a much wider audience than a larger guide would.

Researchers in the UK[64] performed a critical discourse analysis of 25 information security policies within healthcare organisations. They used a methodology to identify truth, legitimacy, sincerity and clarity. In their analysis, they looked for evidence of ambiguity, confusion or lack of explanation, which might ultimately make it difficult for a policy's messages to be clearly and uniformly interpreted by members of staff. It became clear that the there was a significant amount of ambiguity, in particular regarding the policies' objectives and intended targets, as well as significant evidence of the use of jargon and unfamiliar language. Examples of such jargon are "self-regulatory practices" or "best practice" (p.86). The use of obscure and technical jargon could potentially stabilise existing dominant management hierarchies. Furthermore, they found that many of the policy documents are written with an ideological undertone that management has the right to tell other members of the organisation how to behave, to implement surveillance (in order to check on that behaviour) and to sanction those who do not comply. For example, they found that it was common to state that failure to adhere to the policy "may result in disciplinary action or dismissal or lead to involvement of police service" (p.85).

The researchers recommend that policies should locally derived and created, with participation of the largest group of readers and users. It is further recommended that policies use accessible language and terminology and that employees are provided with a separate set of specific guidelines. Concrete examples of issues are helpful to demonstrate the relevance of the policies. Finally, technical content for specialist audiences should be kept in separate documents. This recommendation is supported by researchers that found that staff often feels subjugated by policies[69]. Policies are created without any opportunities for staff to influence the content or to provide feedback. This makes staff feel powerless in the face of sometimes operationally difficult policy directives.

8.3 Internal communication

Internal communication is the continuous exchange of information between professionals within an organization. Information is necessary for employees to perform their tasks (task information), to know the policies, and to know progress (management information), and they need social information about staff issues and social policies. To prevent information underload or overload, information must be timely, correct, relevant, complete, and accessible. Without access to information, employees cannot perform their tasks.

Internal communication to provide employees with information can be:

- Vertical (top-down or bottom-up within a unit)
- Diagonal (between the employee and management of another department)
- Horizontal (between staff at the same level)
- Formal or informal

The communication escalator[70], illustrated in figure 2, describes the intended goals of internal communication. For each goal, there are different communication channels to choose from, depending on the degree of interaction, variety of language, and variety of signals (emotional, social, verbal, and non-verbal) and on the degree of behaviour change intended. The higher on the escalator, the greater the degree of change projected. Awareness is at the bottom, with low interaction with the audience and minimum intention to change. The communication channels used at the awareness level are one-directional, such as email or posters. Commitment is at the top of the escalator, where interaction is high and the degree of change is high. Modern communication channels dominate the top of the escalator, providing many opportunities to create communities that exchange content-rich information and stimulate involvement and commitment of employees.

Communication activities for cybersecurity often remain at the bottom of this escalator. When awareness is the goal, one-directional communication channels such as newsletters or manuals are common practise. To reach the commitment level of the escalator, continuous interaction with end-users is required through collaboration to solve problems and to design solutions.

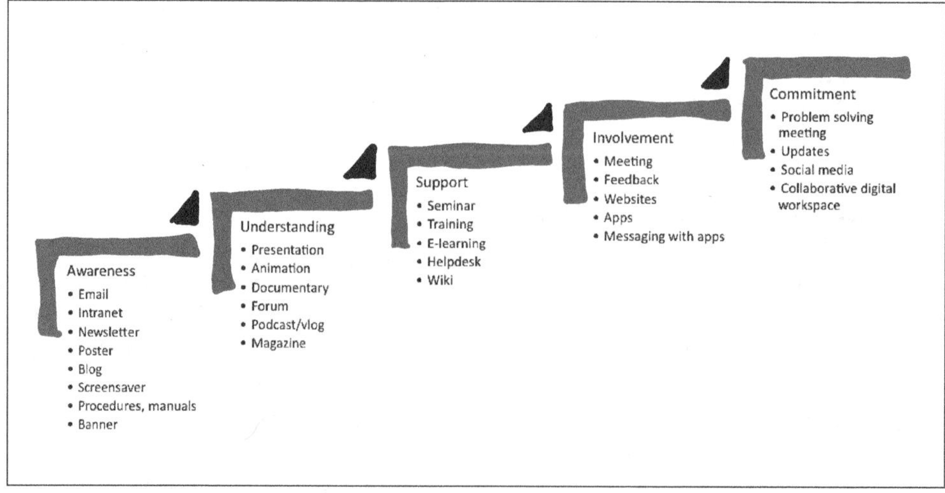

Figure 8 Communication escalator (adapted from Quirke [70])

8.4 Stakeholder theory

Although more than 55 definitions of the word stakeholder exist[71], in general, stakeholders are seen as the individuals, groups, or organizations who can affect or can be affected by a decision, activity, or outcome of a project, programme, portfolio, or the activities of an organization. Stakeholders exist inside and outside of the organization. They are vital for the success or failure of a project, mission, or strategy, and therefore they should be 'managed'. Stakeholder theory states that you cannot look at stakeholders in isolation, each is important for a business to be successful. Figuring out whether their interests go into the same directions is important for success. If you focus on just one group of stakeholders, you miss things; all of them are needed to create success[72]. Many books and academic papers have been written about stakeholder management since the 1990s. They come from different perspectives, including law, marketing, finance, ecology, and industrial relations.

Stakeholder analysis (SA) aims at analysing their relative power and interest, the importance and influence they have, what 'hats' they may wear, and to what networks they belong[73]. For successful engagement, it is necessary to know their stake in the outcomes of your project or strategy. A common tool for SA is a matrix, in which stakeholder groups appear on one axis and a list of criteria or attributes appears on the others. For each overlapping area, a qualitative description or quantitative rating is given. The aim is to gain insight into their position and possible influence on some envisioned end-result; a project goal, conflict solving, organizational change, and so on. A second option for analysing stakeholders is to examine their degree of synergy against their level of antagonism[74]. People with low synergy and moderate antagonism are your opponents; those with high synergy and low antagonism are your unthinking supporters.

Designing a visual aid such as a matrix or stakeholder map during SA helps to understand which stakeholders are the most important. Bourne[75] suggests that some of the dimensions in the picture to be considered include:

Attitude: Will the person help or hinder the work?
Hierarchy: Where is the person in the organization's structure compared to the activity manager: higher/lower, internal/external, colleague or competitor?
Influence: How well connected is the person?
Interest: Does the person have an active interest, passive interest, or no interest?
Legitimacy: Does the person have some level of entitlement to be consulted?
Power: What is the person's ability to cause change?
Proximity: How involved is the person in the work?
Receptiveness: How easy is it to communicate with this person?

Supportiveness: Does the person support or oppose the work?
Urgency: Does the person perceive the work to be important to them?

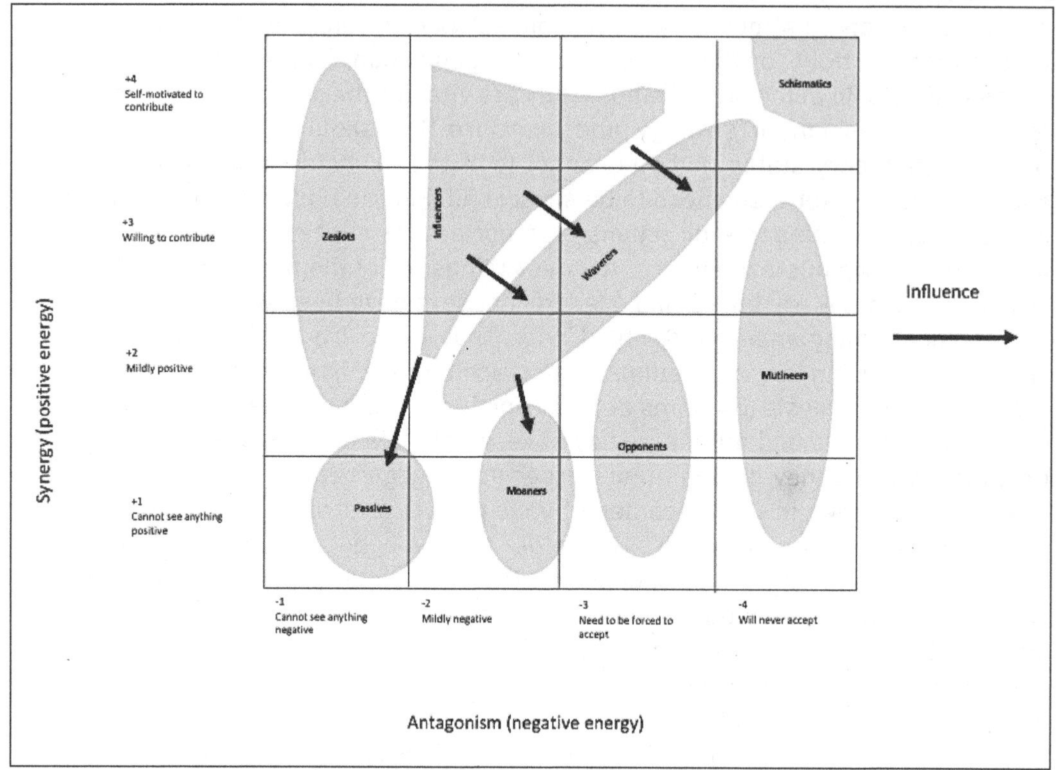

Figure 9 Stakeholder analysis

SA is followed by a communication and engagement plan. A stakeholder communication matrix (SCM) is a tool to plan stakeholder communication activities. It is usually part of a larger communication plan. An SCM may take many shapes but usually includes a list of stakeholders, their interest/power, frequency of communication, choice of communication channels, purpose, and priority. It is meant to keep stakeholders in the loop at the right stages of your activities.

For the cybersecurity professional, it is important to recognize that people and groups inside and outside the workplace influence the success of cybersecurity activities. It may be helpful to identify them, to analyse how they can influence your work, to decide how to maintain a relationship with them, and to decide on the types of communication channels in a stakeholder communication plan. At a final note, different stakeholders may perceive cybersecurity definitions and risks differently, depending on their perspective.

8.5 Communication privacy management theory

Communication privacy management theory (CPM) is about the way people make decisions about revealing and concealing private information[76]. It describes that people control their private information based on personal rules. Individuals believe they own their private information and have the right to control it. They share the information based on their personal privacy rules. When people share information, it means that others are given co-ownership status. As a result, individuals are potentially able to disclose not only their own information but also information that belongs to others. Therefore, the information needs management.

There are three main components to manage private information: privacy ownership, privacy control, and privacy turbulence. *Privacy ownership* means people believe they own their private information and can choose to whom they will deny or provide access to their private information. *Privacy control* represents the management process of providing or denying access to private disclosures. Because the original owner believes they own the information, they also believe that they control access to it even after authorizing co-owners. *Privacy turbulence* occurs when personal boundaries with private information overlap with boundaries for public disclosure.

CPM is used to frame studies on family communication, health (such as studies into wearable devices and sharing of personal health information), social media (sharing personal data in social networks), big data collection, e-commerce, and in studies of the workplace (employee surveillance). CPM has become very applicable in the workplace as personal mobile devices are increasingly allowed at work. The concept of Bring Your Own Device means that employers have to set rules for privacy and security. Companies have to take measures to secure their network. At the same time, some employees feel that companies should not be able to track what is being done on their personal devices or even on company computers even if they are in the work place[77].

A study amongst employees in Taiwan[78] found that employees feel limited privacy in an organizational culture that is controlling. The researchers suggest that organizations better avoid excessive control to improve the negotiation of the privacy boundary between employees and employers. In negotiating privacy with employees, concern about organizational infringement is the main factor influencing trust. They conclude that a good privacy management generates trust in work relationships and encourages a strong motivation and intention of employee commitment and compliance.

In the context of mobile commerce, researchers found that concerns about perceived control boundaries and unauthorized access to personal information have a negative influence on consumers to use mobile commerce. Consumers have fewer concerns about sharing their personal information when they have trust in the commerce partner[79]. Therefore, informing individuals of the extent of data handling as well as clear consent affirmation for future data use is not only a legal requirement under the GDPR, but it also may heighten consumer trust and, as a result, improve the business.

9. How people are persuaded

Persuasion is a form of communication aimed to influence beliefs, values, attitude, intention, motivation, and behaviour. For persuasion, we need a sender, a medium, and a recipient. The sender has a goal and intents to achieve that through communication. The recipient is not persuaded by accident or by force. Persuasion depends on the circumstances in which the recipient resides, the content and form of the message, as well as the credibility of the sender. In cybersecurity, we put a lot of energy into trying to persuade others. The theory discussed in this chapter provides anchors to better understand the processes involved with persuasion.

9.1 Marketing

Marketing is a discipline that sprouted from selling or bartering goods on a market and has since evolved to a science with a focus on delivering added value to customers and maintaining positive relationships. Marketing theory is related to psychology, sociology, economy, social sciences, creative arts, as well as statistics. Marketing theory provides useful insights for cybersecurity when it comes to influencing behaviour. After all, organizations are trying to persuade people to use security procedures, tools, and to act in a certain way.

One of the most traditional marketing models is the AIDA model (attributed to E. St Elmo Lewis, around 1898). The model describes that customers need to go through a number of steps: Attention-Interest-Desire-Action. First you need to draw attention from the people, for instance, by shouting out loud: 'Tasty apples'. Then, the customer must get interested: 'from the best trees, very fresh'. Interest is changed into desire when the customer tastes the apple or sees other people enjoying it. The final step is the actual buying action. This model has since inspired many other marketing models. Today, marketing no longer stops at the buying action but aims at an extended 'customer journey' that includes the post-purchase experience. In modern marketing models, loyalty and advocacy of the product has become the ultimate goal of marketing. Customers nowadays trust the f-factor (friends, family, Facebook fans, and followers) more than they trust marketing communication[80]. People tend to ask their friends on social media for advice. Customers are no longer seen as the 'target' in marketing campaigns. Customers are now considered peers and friends of a brand or product. Advocates tell positive stories and advice their peers. They become evangelists.

Kotler[80] extents the original AIDA to five A's to better fit our modern connected society. The steps are now: aware (I know it), appeal (I like it), ask (I am convinced), act (I buy it), and advocate (I recommend it).

The AIDA model and its newer varieties help us to influence employee behaviour. Organizations that aim for security awareness programmes as a checkbox exercise for compliance to standards and regulations do not have do to more than expose a target group to their messages and training courses. When we aim for employees to act, and re-act, such as periodically changing a password, we should take employees through all the stages from awareness to action. However, organizations that aim for structural changes and long-lasting behaviour need to take the extra steps to achieve employee loyalty and advocacy of secure behaviour. Several studies found evidence that observing the secure behaviour of peers motivates the secure behaviour of followers[11,13,81]. To change security behaviour, we need advocates in the peer groups of our audience.

Researchers[65] who studied the effectiveness of sanctions on information security violations found that a policy enforcing strategy has limited effect. They found that enforcing becomes non-significant when an influencing strategy is used. Employees can be educated to hold themselves accountable for their actions. The authors suggest that there should be a focus on the link between employees' actions and business risks. Another tactic is to use role models who understand security issues and help to advocate policy compliant behaviour. However, the same researchers furthermore discovered that the more senior the position of an employee, the more likely this person is to violate security rules. This conclusion points out an interesting topic for future research, as most best practices state that security governance ought to be driven by senior executives.

9.2 Rhetorical theory, framing, and narrative paradigm/ storytelling

Classical rhetorical theories were dominated by the ideas of Aristotle and Plato with a focus on persuasion through language and speech in politics and law. In the classical theory, there are three modes of persuasion:

Ethos: to convince the audience of the credibility of the speaker. This is done by choosing the appropriate language, clothing, and by demonstrating expertise
Pathos: to persuade the audience by appealing to their emotions
Logos: the appeal to logic to convince the audience

Contemporary rhetorical theory now addresses not only speech but all contexts in which symbol use occurs including personal diaries, television, the Internet, and websites as rhetorical artefacts. This means that rhetorical theory also includes the study of visual and non-verbal elements, such as the study of art and architecture, buildings, cars, appearance, sports, and so on. Visual rhetoric can be embedded in a performance, body language, pictures, objects, physical spaces, and so on. For example, a picture of a medical doctor often shows the doctor in a white coat, to make the credibility of the image stronger.

We can recognize *verbal* rhetoric in cybersecurity in stories about cyber-crime, cyber terrorism, hacktivism, or cyber warfare. The growth of data breaches and malware has brought these stories to the front pages of the media. These stories have impact on the public perception of cybersecurity. Reporters find writing about cybersecurity challenging: selling the story to the editors requires public interest which is sometimes lagging, and they have to persuade organizations to share information about data breaches[82].

Stories are a powerful tool to explain security goals to developers and C-level executives[83]. Through a story or scenario, people can better visualize the importance of the message or the risk. However, the story must be relevant to the organization and supported by some analysis[82]. This was already recognized by the founder of the concept of the narrative paradigm, Walter Fisher, in the 1980s. His theory claims that stories are more persuasive than arguments alone. The concept supports the idea that communication and persuasion is effective only when it makes sense to the listener. Therefore, the story must be coherent (structured and contain credible characters) and credible (factual events, reasoning pattern, and importance).

Some academics state that stories told by cybersecurity specialists are exaggerated and oversimplified[84]. The possible effect of framing stories in cyber-doom scenarios is even counterproductive: the use of extreme fear can lead to a sense of fatalism and demotivation to act[85,86]. Framing is a technique to present a message in a positive or negative context, to reach the maximum effect with the audience. For instance, when respondents are asked whether they support a group of extremists in terms of freedom of expression, most will answer positively. But when they are asked to support the same group framed in terms of violence, most people will oppose. Message framing requires making the message less complex and easy to explain. In cybersecurity, this is often a challenge. De Bruijn and Janssen[86] recommend six framing strategies for cybersecurity:

1. Do not exacerbate cybersecurity. Keep it realistic.
2. Make it clear who the villains are.
3. Give cybersecurity a face by putting the heroes in the spotlight.
4. Show its importance for society.
5. Personalize for easy recognition by the public.
6. Connect to undercurrent. Cybersecurity is interwoven with other issues.

In 2018, the Dutch government released the news that foreign state spies were caught hacking[87]. During the press conference, the government provided a detailed timeline of events, supported with images of the people involved, details from passports, equipment, and a photograph of the accused people at the airport. It was an unusual sight; normally, the government does not share details of these intelligence operations. The framing of this message followed all six of the above strategies and formed a clearly framed message to the involved state actor that the Dutch government finds these activities unacceptable, supported by visual rhetoric and stories.

Cybersecurity professionals can improve the persuasiveness of their communication by using the key takeaways from rhetoric. For instance, by telling a story that involves a credible hero, a villain, and an appeal to the emotions of the audience. With a clear message that the audience can relate to ('this could happen to me') or that matches stories people already know ('that is just like what was in the news the other day'), the audience makes a mental picture of the situation and are likely to internalize the message.

An example of a story that involves a hero and an appeal to emotions of the audience is visible in a video message from Kaspersky Lab Benelux[88], responding to the decision of the Dutch government to stop using Kaspersky's software. This video shows a likeable hero explaining all the good work and showing the sympathetic team that works in The Netherlands. This video uses visual rhetoric and satire to influence public trust in the brand.

9.3 Elaboration likelihood model

The elaboration likelihood model (ELM) is based on the idea that attitude change follows one of the two basic mental routes. They are known as the two routes of persuasion. The **central** route of persuasion is the logical, conscious, and thoughtful route. In this route, the person thinks about a message for a longer duration of time and carefully considers the information that is given. The process of thinking for a longer duration is known as elaboration. To achieve this, people need motivation: they should be interested in the information. If the persuasion level of the message is strong, well-constructed, convincing, or creative, people will get persuaded to change their attitude and behaviour. Attitudes obtained via the central route last longer, are less vulnerable to contra-argumentation, and are better predictors of human behaviour. The **peripheral** route is used when the message recipient has little or no interest in the subject or has a lesser ability to process the message, or when the content is weak. With the peripheral route, people do not pay attention to persuasive arguments and are instead influenced by superficial characteristics, such as the popularity of the speaker, and the change in their attitude is consequently temporary.

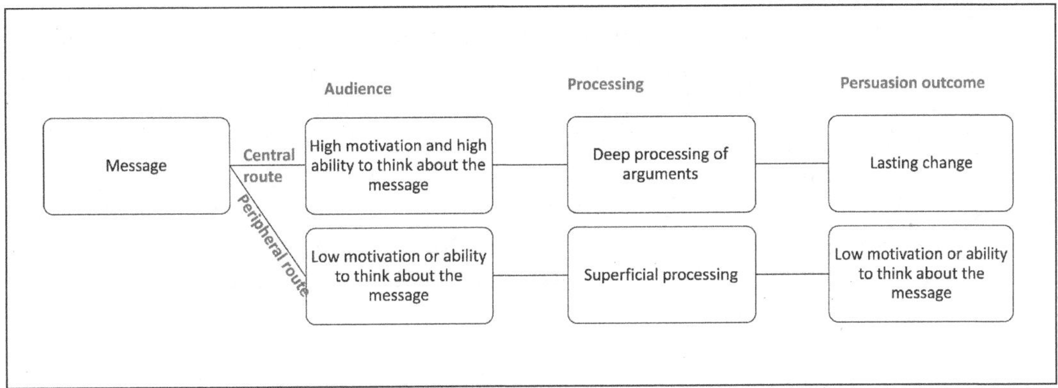

Figure 10 Elaboration likelihood model.

The model is useful when planning communication for attitude change:

- Elaboration requires concentration and distraction will disrupt the process.
- Repetition of messages is likely to increase the possibility of elaboration, but too much repetition eventually leads to the peripheral route.
- When the audience is motivated to elaborate a message, rely on facts and strong arguments. Weak arguments will backfire and cause discussion.
- If the audience is unable or unwilling to elaborate a message, make the presentation more important than the message. Appealing messages will be processed on the peripheral route. However, the level of persuasion will be fragile.

Researchers have tested the ELM in the context of security awareness. Researchers in Singapore tested a security message from a security awareness campaign of a university, informing students about an anti-virus software package that they could install on their computers[89]. Four versions of the security message were created, i.e., high/low on message comprehensibility and with/without message repetition. They found that repetition of a security message enhances the elaboration likelihood. Message comprehensibility only had a positive effect on elaboration when used in combination with repetition. They argue that including some information security jargon in a message can be reassuring to readers and give precise meanings.

In Finland, a study used the ELM as underlying theory for information systems security policy compliance training[90]. They found that security policy compliance training should use learning tasks that are of personal relevance to the learners so that there are visible consequences for the self and others. They found that employees' long-lasting learning is enabled by having them exploring the unwanted consequences of their own actions. They also found empirical evidence that one-off training is not enough, and that training should be supported with continuous communication activities.

A study amongst 325 students in an American university[91] tested individual responses to phishing emails and found results consistent with ELM and other related theories. The study found that several variables influence the elaboration likelihood:

1. Level of involvement in the phishing detection process. The more relevant an email is, the more elaboration.
2. Attention to urgency cues. Phishers use urgency cues to communicate fear or threat. In this study, the individuals focused disproportionately on urgency cues, often ignoring other elements of the email such as its source and the grammar and spelling used in the email. The lack of attention to these elements increases the individual's likelihood to be phished.
3. Habitual media use patterns, where individuals inattentively respond to relevant emails, accounted for at least one-half of the phishing susceptibility.
4. Email load. Individuals were far more likely to respond to phishing emails in the presence of large email loads.

The main lesson is that, according to the ELM, people are more likely to be persuaded if they engage with a message longer and repeatedly. To achieve longer attention, the message must be understandable and personally relevant. Distractions should be avoided.

9.4 Cognitive dissonance theory

Cognitive dissonance is the discomfort that people feel when they simultaneously hold conflicting beliefs, ideas, or behaviours. For example, when a person believes it is important to exercise but rarely makes time for sports. This causes a feeling of discomfort that may stimulate the person to change the behaviour and go to the gym. Other ways of dealing with the discomfort are justification of their behaviour by changing attitudes (I do not have to go: the health benefits are overrated, I get injuries) or by adding more arguments to their beliefs (it is expensive, I am too busy).

Knowledge about the particular cognitive dissonance of a person gives the potential to amend that person's behaviour. This makes it a popular theory amongst advertisers, salespeople, activists, and politicians. If marketeers want people to buy their product, they feed consumers with information that forces them to make a decision. Think, for example, of commercials for health-related products. They argue that a regular product is not good enough; your symptoms will not go away with the other products, but when you buy their product you can enjoy life again.

There are several theories about reducing dissonance between attitudes and actions:

- Selective exposure. People tend to select information that fits with what they already believe and ignore the opposite. A warm and personal environment is best for considering different views.
- Need for reassurance after a decision. Dissonance is greater after an important decision was made, especially if it is hard to change it back and if it took a long time to make the decision.
- Minimal justification. The best way to change attitude is to offer just enough incentive. If there is no significant reward, people will change their attitude to justify their behaviour. If there is a large reward, people will achieve compliance without attitude change.

Steps to reduce dissonance of end-users regarding cyber secure behaviour are stimulating new beliefs to change the weight of the conflict (e.g. by demonstrating how easy it is for third parties to access their computer), to provide them with new information (show them solutions such as password management software), or to reduce the importance of the beliefs (by enforcing certain actions through technology).

Cognitive dissonance is a phenomenon that is targeted by cybercriminals. A study amongst students in South Africa found that students were confident about their attitude and skills in security, their actual behaviour showed differently, indicating that they will behave in a way they know is wrong[7]. For instance, they were trusting email content sent from email accounts with names of their friends, even though they were confident about their skills in recognizing phishing. Similar results were found in the attitude versus cyberbullying and perception of skills for avoiding cyberbullying. Finally, the students knew that complex passwords are more secure but choose simple ones anyway. These examples of cognitive dissonance make the students vulnerable for cybercrimes.

Another example is the privacy paradox. Online users claim to be very concerned about privacy, but at the same time, take little action to protect their personal data. Several studies have tried to explain the rationalization that people use to explain their behaviour but with mixed results[92–94].

Another example is password fatigue, which occurs when users are faced with an overabundance of passwords to remember. Ninety-one percent of respondents in a survey amongst 2,000 adults around the world[95] said that there is inherent risk associated with reusing passwords, yet 61% continue to use the same or similar passwords anyway, with more than half (55%) doing so while fully understanding the risk.

9.5 Nudge theory

The nudge theory is often used in politics, healthcare, marketing, and organizational communication to influence behaviour. A nudge is a small change to the context within which a decision is made. A nudge, as defined by Thaler and Sunstein[96], is any aspect of the choice architecture (the set of options to choose from) that alters people's behaviour in a predictable way without forbidding any options or significantly changing their economic incentives. Changing the choice architecture makes it easier for people to make decisions that are in their self-interest. A famous example is the fly that is painted on urinals in restrooms. Men tend to aim at the fly, thus reducing spillage and cleaning costs. It is not mandatory to aim at the fly and there is no specific economic benefit for the visitor of the restroom. They could still choose to aim somewhere else. Another well-known nudge is to set a default option as opt-in for specific choices that are better for the community: opt-in as default for organ donation or pension-saving schemes.

Nudges are frequently used in a cybersecurity context to influence behaviour of an end-user. Nudges for security and privacy face some challenges regarding ethics, criminal exploitation, and design. There are concerns about the impact on the nudgees' welfare, autonomy, and dignity[97]. Nudging ought to be carried out for the good, but the question is, whose good? There is ongoing debate about the ethics of nudging, especially when carried out by governments. The danger with nudges is that people are forced into political or religious directions. Most people tend to support nudges as long as they serve to promote health, safety, and environmental protection[98]. Sometimes, however, they are based on lies and manipulation. Nudges can be used by cybercriminals to nudge people into disclosing personal information, to install malicious software, or to click on something that looks attractive but eventually leads to identity theft or fraud. Another criticism is that design choices may lead people to start ignoring risks. They may rely completely on the default settings and stop thinking about security.

The table lists some examples of nudges in cybersecurity.

Table 3 Overview of nudges used in cybersecurity.

Desired behaviour	Nudge to stimulate a better decision
Choosing strong passwords	Messages and password strength meters encourage users to choose stronger passwords or stronger security questions.
Changing default settings to opt out	Design software to pre-tick the box that offers users to opt out of using their personal data. Internet and mobile app users hardly change default security and privacy settings. Changing the default settings to 'do not send me newsletters'[99] hardly leads to users choosing the opposite, less privacy-preserving, opt-in option.
Keeping personal data private	Showing people daily how often smartphone applications access their personal data to motivate users to update their privacy settings.
Activating privacy settings	Some social media websites show icons or messages that show the user who can see and share what they post and how to change those settings.
Installing updates	Notifications to install updates or installing an update unless the user postpones it by clicking on something.
Safe online shopping	Warning messages to remind consumers to navigate safely at the beginning of a shopping exercise to explain how to connect securely, how to find trusted vendors, and to log out after finishing[100].
Use screen lock when inactive	Automatic screen lock after a period of inactivity.
Using reliable Wi-Fi networks	Ordering or colour coding the list of available networks with the most secure ones on top[101].

9.6 Theory of planned behaviour

The theory of planned behaviour (TPB) is used to predict a person's intention to engage in a behaviour at a specific time and place. The theory states that to predict what people are going to do, we need to know their intentions. Intentions are influenced by attitude, norms, and perceived behavioural control[102]. Attitude refers to the degree to which a person has a positive or negative evaluation of the behaviour. Norms refer to subjective and social norms, for instance, the individual's perception about a behaviour is influenced by the judgement of parents, friends, teachers, and so on. Perceived behavioural control refers to the individual's perception of his/her own capacity to carry out particular behaviours.

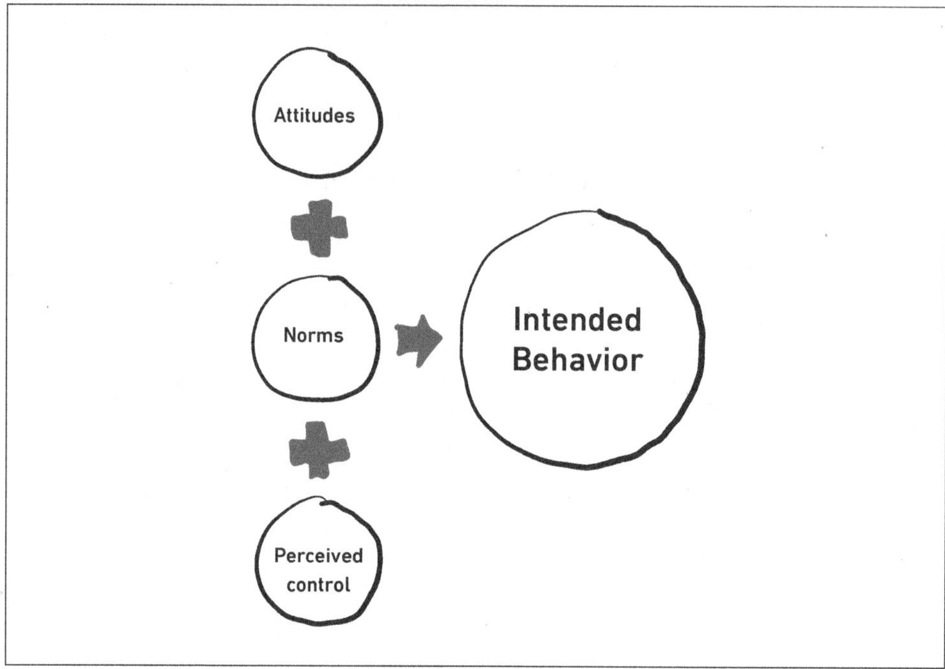

Figure 11 Theory of planned behaviour.

A small number of researchers have used the TPB in relation to cybersecurity. Australian researchers investigated online safety behaviour amongst users in different countries and found that perceived behavioural control was the strongest predictor of intended behaviour[12]. Furthermore, social norms are also a strong predictor. This suggests that influence from the social group that a person belongs to has a significant bearing on whether an individual intends to engage, or not engage, in secure behaviours. This result matches with visions from modern marketing theory where customers are seen as advocates of a brand or product.

10. Message presentation

10.1 Gestalt principles

The Gestalt school of thought started around 1912. More than a century later, Gestalt psychology continues to inspire research into vision science and cognitive neuroscience[103]. However, most people will associate Gestalt with design and arts. Gestalt psychology forms the foundation of design rules still used today. It refers to perception of grouping: people, when looking at an object, first see the whole of something and after that, they start seeing the details. This process is called emergence. When we see an object, we first seek for its outline. Then we match this outline against shapes and entities that we know. After that, we start identifying the parts that make up the whole object.

Another process is called reification. If we cannot find a pattern match to an object, we search our memory to find something that looks like it and we fill in the gaps. Some pictures can be interpreted in multiple ways. Our mind then chooses to see one at the time.

There are several principles that help to understand what people observe and thus help to design a visual aid in a way that it will look coherent and complete. They are called the principles of grouping or the laws of Gestalt.

Table 4 Overview of Gestalt principles

Principle of grouping	Effect	Example
Similarity	The most similar elements (colour, size, texture, type font, or orientation) are perceived as more related than elements that do not share those characteristics.	
Closure	We tend to look for a recognizable pattern. We fill in the missing information to complete a figure.	

Proximity	Objects that are closer together are perceived as more related than objects that are further apart.	
Figure/ground	Elements are perceived as either figure (the element in focus) or ground (the background on which the figure rests).	
Parallelism	Elements that are parallel to each other are seen as more related than elements not parallel to each other.	
Symmetry and order	People tend to perceive objects as symmetrical shapes that form around their centre.	 < >< >< >
Continuity	Elements arranged on a line or curve are perceived as more related than elements not on the line or curve.	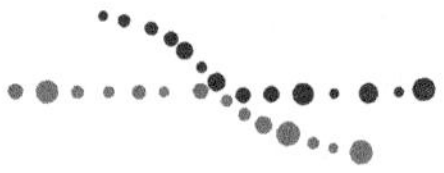
Common region/element connectedness	Elements are perceived as part of a group if they are located within the same closed region.	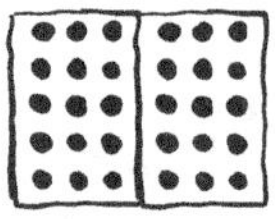
Good figure/ Prägnanz/ simplicity	People will perceive and interpret complex images as the simplest form(s) possible. The picture on the right is easier to read as three shapes than as one complex shape.	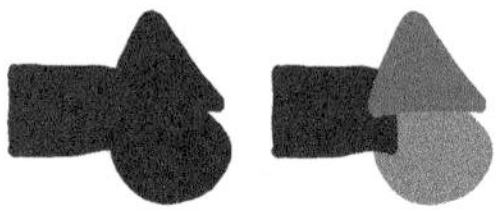
Synchrony/ common fate	Elements that move in the same direction are perceived as more related than elements that are stationary or that move in different directions.	
Uniform connectedness	Elements that are visually connected are perceived as more related than elements with no connection.	
Focal points	Elements with a point of interest, emphasis, or difference will capture and hold the viewer's attention.	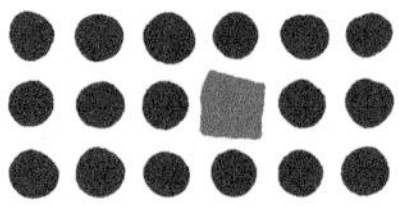

The principles can be used to improve the composition in a presentation or on the design of a poster, a warning message, website, or infographics. Being aware of and implementing the principles of Gestalt theory can have the effect of making visualizations feel stronger and more coherent.

Gestalt can also be applied to data visualization. Data visualizations for cybersecurity are often created for IT experts and analysts (e.g. for their work with log files, malware, and vulnerabilities) and show common bar charts, pie charts, donuts, circles, heatmaps, and tables. However, data visualizations with the non-expert end-users in mind may benefit from Gestalt principles. For instance, researchers in New Zealand[104] designed a user-centric visualization based on the principles of proximity, similarity, Prägnanz, and common fate for end-users interested in the life cycle of their data in the cloud. They aimed to find techniques to transform raw data (logs) into a final output that end-users could easily understand and assess for security risks. The inclusion of the Gestalt's theory of perception to the technique provided a range of visualizations that allowed the end-users to visualize and interact to track data provenance.

Bad actors also make use of Gestalt principles to set up fraudulent documents. For example, a small change in a corporate logo on a fake document may be overlooked because of the human perception of that logo. If it looks familiar, a small change in a detail is likely to be overlooked. Current approaches in security and forensics analyse patterns in images and Gestalt principles are used to improve pattern recognition in security and forensics analysis[105]. Examples are the detection of digital watermarks, recognition of spam email, breaking cryptographic schemes, analysis of fingerprints, or digital forensics.

10.2 Repetition and habituation

Habituation is a form of learning where the response to a stimulus decreases after being repeatedly exposed to it. People habituate and, as a result, respond less to repeated security warnings over time. Neurobiologists argue that this is not something people do on purpose but that it is a natural consequence of how the brain works. Habituation is an important survival mechanism because it allows organisms to save their energy for relevant threats.

For cybersecurity, this implies that after the first exposure to a security warning, the attention to this warning drops when it is repeated over time. The attention can improve again by improving the design of the warning messages, by frequently changing the design (border, colour, size, symbols, and text) over time, and by giving the users a temporary break from the messages before starting again.

In an experiment with eye tracking technology, researchers compared the user's attention to conventional warning messages with warning messages that were designed to change their appearance (polymorphic warnings)[106].

The polymorphic messages comprised nine graphical variations as advised in warning science literature, including the colour red, pictorial symbols, signal words, contrast, and ordering of options. The results demonstrated that the polymorphic warning is substantially more resistant to habituation than static warnings. Further research with functional MRI scanning of the brain provided even more evidence that users habituate to warnings within a few days but habituate less to polymorphic warnings[107]. The use of colours also influences attraction. For instance, secure sockets layer or password warning messages on websites perform better in red than in yellow[108,109]. Thus, messages that warn users for privacy and security risks should vary in design every time they appear on screen and the designer of the message should consider colour variations.

10.3 Semiotics

Semiotics is the study of signs and signification. Signs represent objects that are not physically there. A sign can be a word, an image, a performance, a multimedia production, a building, or even a trend. Also sounds, fashion, and traffic signals could be signs. Signification is the process of creating meaning to a sign. Signification is related to the way a sign is encoded, transmitted, and decoded. For people to understand the meaning of a sign, they must be able to encode and decode it. The decision that a word, colour, or symbol is a code for something is made by the social group (culture), as they agree on the meaning of it. Take, for example, some specialized vocabularies (work, office, and sports), various kinds of slang, street art, or the use of emojis. We only know what the words or symbols used there mean if we belong to the group and learn using one of these kinds and styles of language. Semiotics is a learned and a shared conceptual connection in all uses of signs (language as well as visual).

There are mainly three types of signs: (1) an icon, which resembles its referent (such as a road sign for falling rocks or pictograms); (2) an index, which is associated with its referent (as smoke is a sign of fire, a fingerprint relates to a person); and (3) a symbol, which is related to its referent only by convention (as with words or logos). A symbolic relation between an object and a sign is more abstract and can only be learned through learning the cultural codes in a society.

There is a growing amount of activities using semiotic theory related to privacy and data protection. There are several initiatives to make law more accessible, more useable, and more engaging through legal design[110–113]. The field of legal design promotes that new methods of communication should be considered to allow any individual, even a lay person, to access and understand legal information. In the case of privacy, this is even mandated by law. Article 12 of the GDPR requires the provision of intelligible and easily accessi-

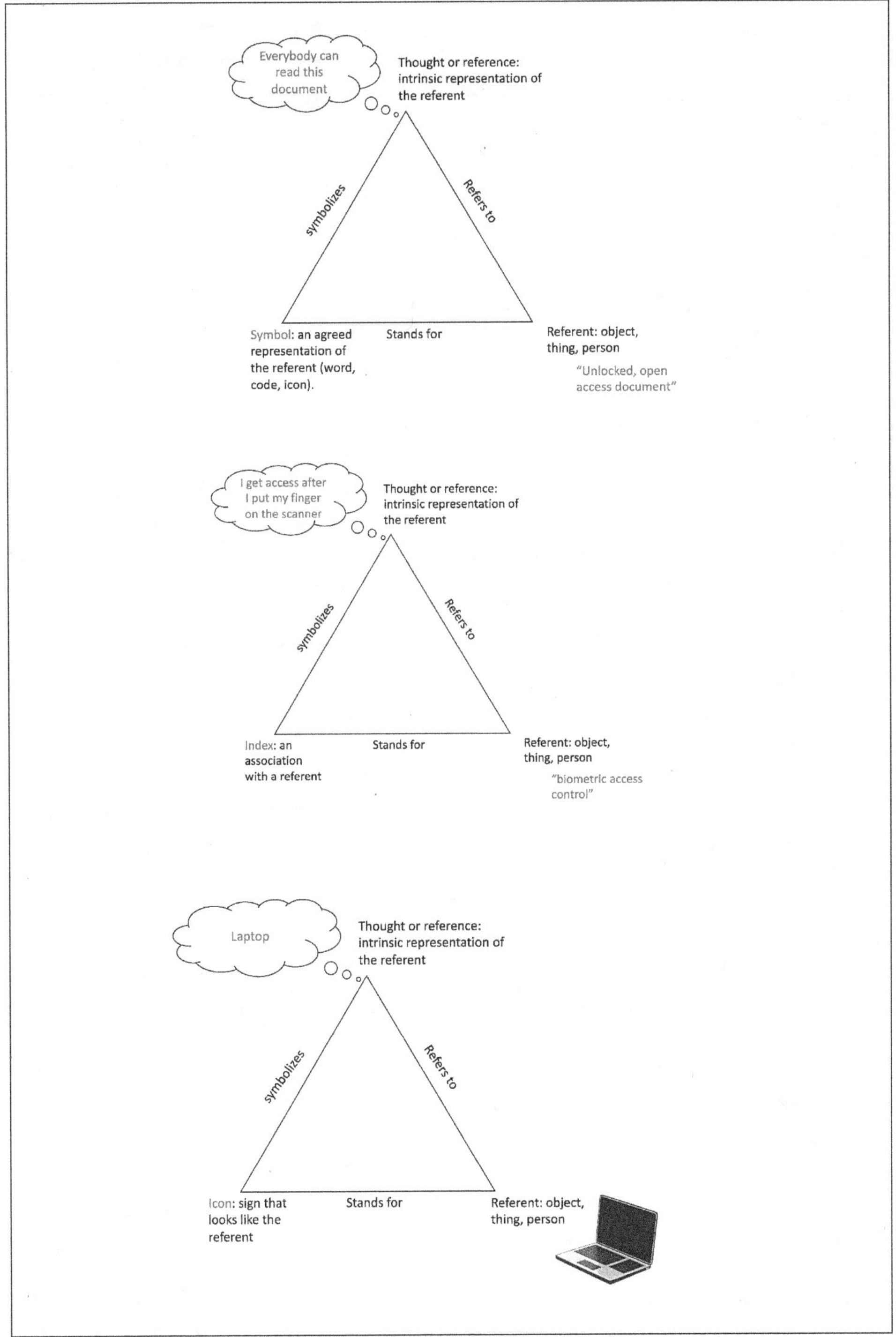

Figure 12 Overview of semiotic triangles. Three semiotic triangles explaining icon (recognizable picture), symbol (meaning of the picture needs to be learned), and index (association with the object or idea, they do not look like each other).

ble information on data practices. Article 12(7) even suggests the use of icons 'in order to give in an easily visible, intelligible and clearly legible manner a meaningful overview of the intended processing'. Icons can represent data protection concepts and their interpretation and function as a shared language. When design, communication, and information technology collaborate, it is possible to generate user-friendly interfaces to the law[114].

10.4 Personal branding

Personal branding is a marketing strategy that a person adopts in order to promote his or her major personal characteristics. Personal branding is common amongst politicians, actors, manager, scientists, and entrepreneurs. With social media, everyone can be a broadcaster and can create a personal brand. Personal branding became popular in the late 1990s under the influence of business articles, such as The Brand Called You by Tom Peters in 1997 and Managing Oneself by Peter Drucker in 1999. The discourse sees mainly publications in the form of self-help books and career advice and is often criticized by academia as a fantasy. Some state that personal branding sells hope instead of the person: 'the hope of standing out, the hope of being acknowledged, feeling unique and worthy of attention and most of all, the hope of finding meaning'[115(p6)]. Academic papers that discuss personal branding sprout mainly from the fields of sociology[116] and marketing[115].

Most people already subconsciously market themselves and they think that personal branding is effective. For instance, when job seekers are interviewed, most of them have internalized the concept of personal branding and are very conscious about their online presence and image[116]. Sales people and executives in a diverse set of companies in Europe and the USA responded in interviews that they felt that their personal branding provided them with tangible results (e.g. sales results, customers calling back, awards, and trips) and intangible rewards (clients smiling at you, getting compliments, etc.)[117].

A growing number of cybersecurity experts are promoting their personal brand. These people are seen as top influencers in cybersecurity. This is not only important to promote cybersecurity as interesting field for recruitment purposes or to influence and advocate secure behaviour, but it also helps to advance one's own career. Jane Frankland, a business woman in cybersecurity, describes in her book[118] how she decided to improve her own personal brand in cybersecurity, invested time to learn how to use social media, and reaps the benefits from it ever since. Other influencers became famous because of their actions (Kevin Mitnick), knowledge and writing (Bruce Schneier), or discoveries (Dan Kaminsky). They are seen as role model and influencer. Personal branding helps to gain your audience's trust as your reputation as a knowledgeable expert becomes more visible. By showing your knowledge or relation with a reputable organization, you will be viewed as someone with authority[3]. It will help you build relationships and gain credibility.

Personal branding is largely, but not only, about your online presence. It requires visual communication. Examples are storytelling to deliver your message, an attention-grabbing biography on a website, a catchy LinkedIn headline, your personal style and design in presentations, professional profile pictures, as well as video messages. Most influencers have a visually attractive website and a consistent visual style in articles, blog, and presentations. They know what they are talking about and stand by their norms and values. They live by those rules all the time and are always conscious about others seeing them.

11. Conclusion

Communication theory guides us to make better communication design decision. It provides models to understand the context in which communication takes place so that we understand our audience better.

It also demonstrates how people learn and how they receive and consider security messages. Communication research demonstrates that remembering and understanding information does not necessarily lead to commitment and advocacy. To set up a communication strategy, there are many choices such as using marketing models, rhetoric, storytelling, framing, or personal branding.

The key takeaways from communication theory are:

1. Tailor security education to the cognitive level of learning objectives
2. Mix visuals, spoken words, written words, and hands-on for maximum effect
3. Adjust communication style to match the organizational structure
4. A shared culture = a shared language
5. Stakeholders influence success
6. Trust in employer motivates compliant behaviour
7. Messages must be relevant and repeated
8. A visual has more effect when the design is coherent
9. Information should fit with what is already believed
10. Persuasion need trust and reassurance
11. The strongest influencer is the social group that a person belongs to

In the next section of this book, you will find many ideas to support your communication strategy. A selection of these examples is briefly discussed in relation to communication theory, to help you train your mindset when you are looking at pictures.

Part III

Visual communication in the CISO office

12. Introduction

A Chief Information Security Officer (CISO) and her team have 1001 tasks to do. Visual communication can help to work more effectively and to engage stakeholders. This part of the book suggests some common visualizations for the different concepts that the CISO team may be explaining to different audiences. Remember that the examples in this section are only the tip of the metaphorical iceberg. Our community has only just begun to realise the power of visualization and this part of the book is only the beginning of a long and creative journey. Other fields such as legal design, information design, and data science are important innovators that cybersecurity experts should collaborate with. Much work is already coming together at events such as the IEEE Symposium on Visualization for Cyber Security (VizSec), running since 2004[119], and the annual International Workshop on Graphical Models for Security (GraMSec) since 2014[120]. Raffael Marty stated in his book Applied Security Visualization (2008):

'A picture is worth a thousand log records'[121(Page xiii)].

In his vision, visualization is the process of generating graphs based on data. In the examples above there is a strong focus on data visualization and tools rather than communicating complex concepts to non-experts.

Other work creates security from visual information. For instance by taking pictures of a moving freeway, clouds, or lava lamps to create randomness for cryptography[122].

In this book, visualization is primarily about communicating information in the clearest way possible, considering the target audience and the context. The examples in the next chapters demonstrate how to use visualization in different contexts of cybersecurity. Our eyes cannot see cybersecurity risks, threats or organizational structures, but when presented in visual form, these concepts are easier to understand. The examples all show different styles, techniques, and messages. Some originate from commercial sources, others come from public sector and academic publications. Hopefully, further editions of this book and a supporting website will allow our community to build up a library of examples that can inspire future visual communicators.

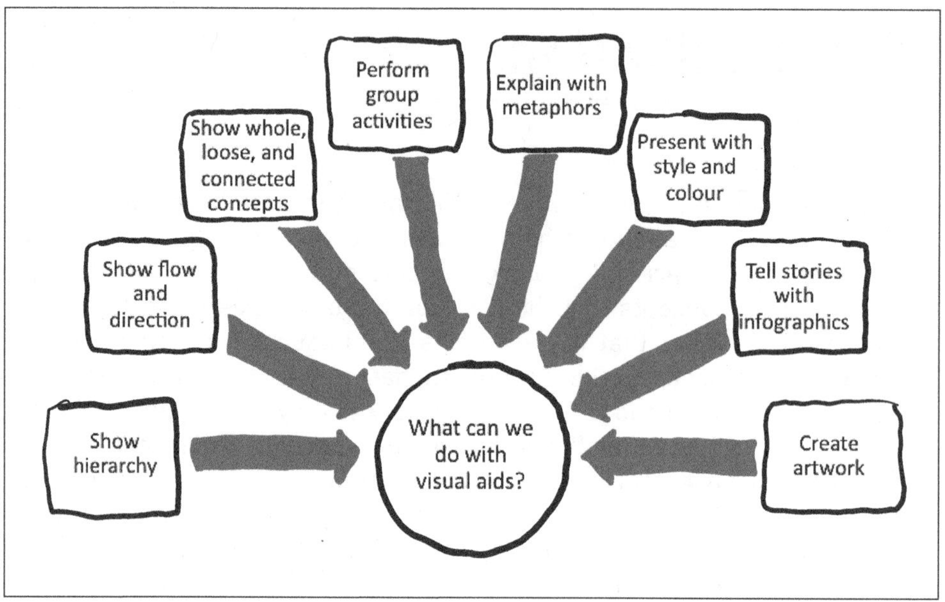

Figure 13 How can visualization help?

13. Visualizing hierarchy

Hierarchy is typically visualized in tree diagrams. A tree can be used to display textual information as well as numerical data. Some examples of situations where a tree visualization is helpful are:

- To display a taxonomy
- To show family relations and descent
- To show decision-making options
- To calculate probabilities
- To show an organizational structure

There are many forms to display a tree. The basic shapes are horizontal or vertical trees, rectangular tree maps, circular tree maps, and radial trees.

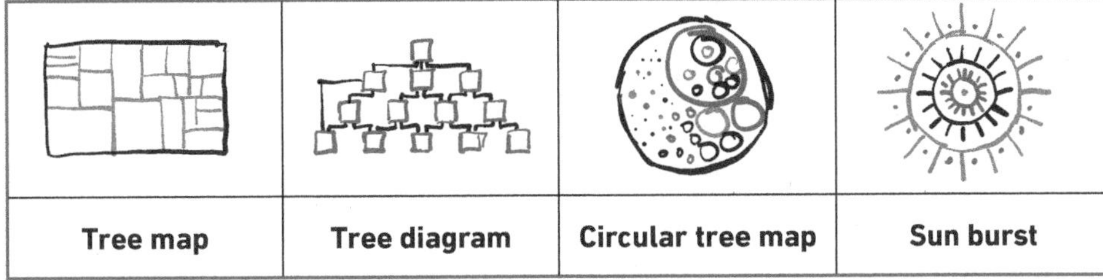

Visualization of data in tree diagrams is one of the simplest forms of presenting data. Many examples of these types of hierarchy visualisations can be found in academic literature and within a business context. Some references to examples are:

Tree map: tree map of vulnerability categories[123], tree map of trojan families[124(p12)]
Circular tree map: status of tokens and services across subnets[125], active nodes in software-defined networking[126]
Fault tree/attack tree diagrams: CORAS model for risk analysis[127], sensor cloud framework risk assessment[128], CVSS attack graphs[129]
Sunburst: Mitre ATT&CK framework[130]
Multi-directional tree: patterns of breaches[131], attack reactability[130]

Tree map

Treemap of actors at the start of an information incident scenario and their location.

Nicole van Deursen

This treemap is based on data from research in healthcare organisations between 2006-2010[25]. It shows that the majority of investigated incidents (n=2108) within healthcare organisations started with an internal employee (74%), mostly within the premises of the organisation, but also sometimes at home or at the patient's home. Other main starting points for incidents are business partners (2%), software/end-user applications (5%), and many incidents (18%) have unknown causes.

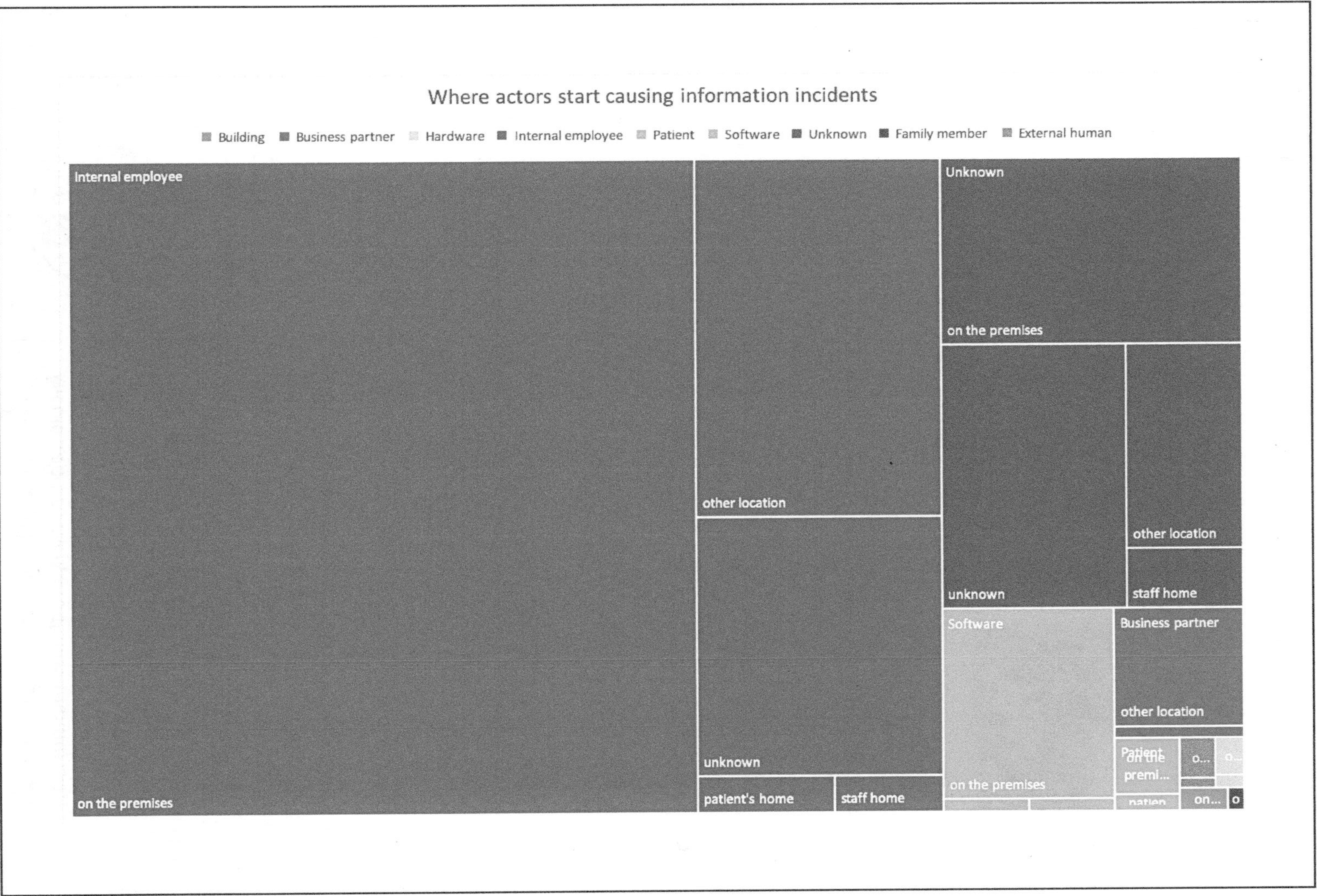
Where actors start causing information incidents
Building
Business partner
Hardware
Internal employee
Patient
Software
Unknown
Family member
External human
Internal employee
on the premises
other location
unknown
patient's home
staff home
Unknown
on the premises
unknown
other location
staff home
Software
on the premises
Business partner
other location
Patient
premi...
o...
on...
o

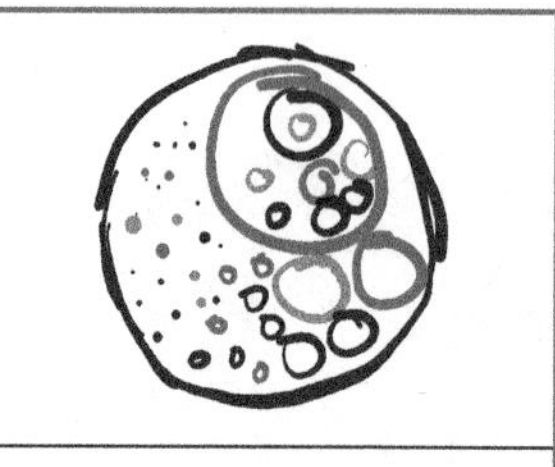

Circular tree map

A taxonomy of information risks in Healthcare
Nicole van Deursen, created with Flourish
Based on: [25]

A taxonomy is a system for naming and organising things into groups that share similar characteristics. Taxonomies are often organised hierarchically, with different levels of groups and sub-groups. This visualization of a taxonomy of information risks in healthcare shows the categories (threat, method, weakness, event and damage) that together describe a risk scenario, and their sub-categories.

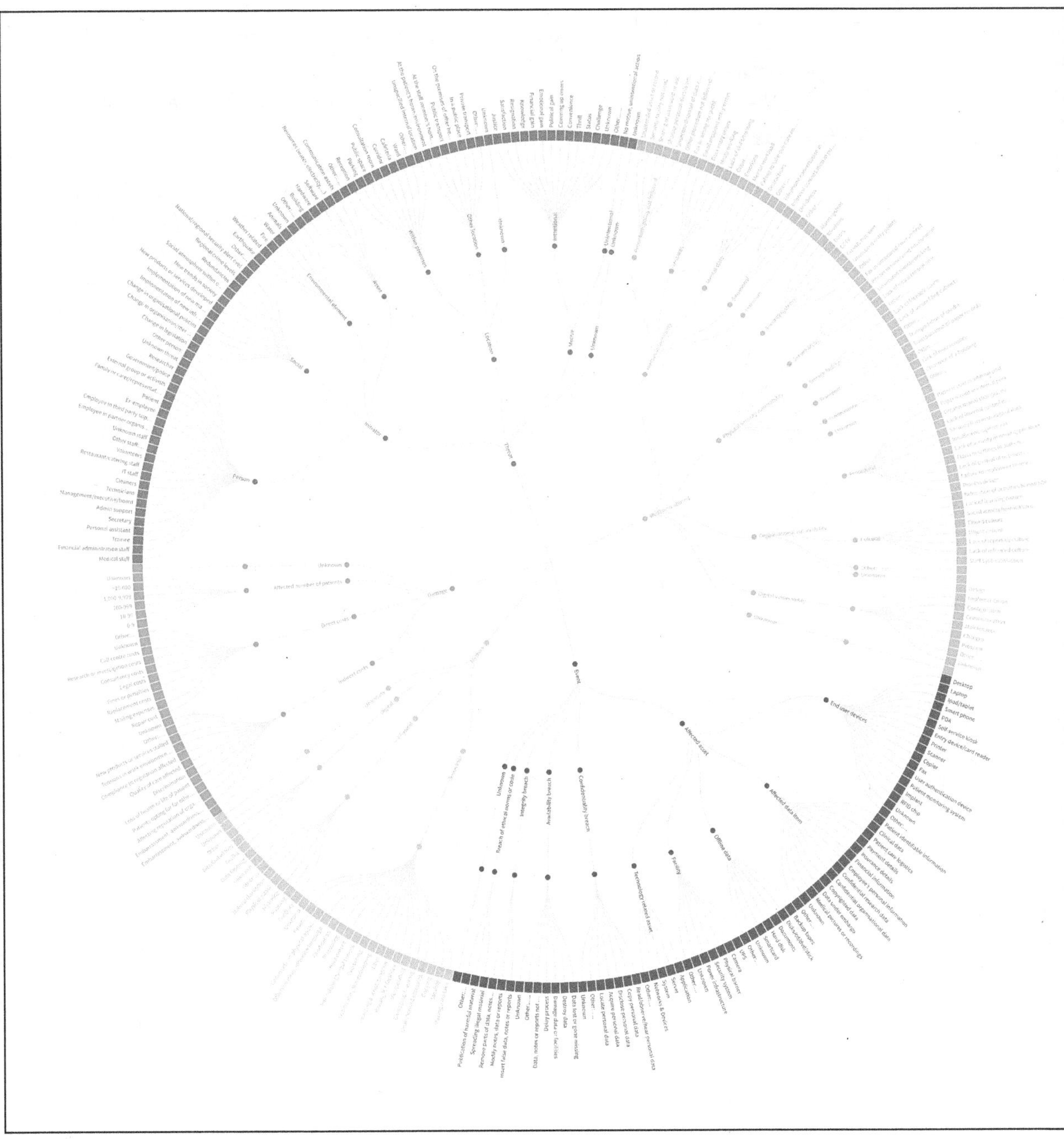
Threat
Event
Initiator
Person
Social
Environmental element
Location
Motive
Affected asset
Affected data item
End user devices
Offline data
Facility
Technology related asset
Confidentiality breach
Availability breach
Integrity breach
Breach of ethical norms or code
Unknown
Desktop
Laptop
Ipad/tablet
Smart phone
PDA
Self service kiosk
Entry device/card reader
Printer
Scanner
Copier
Fax
User authentication device
Patient monitoring system
Implant
RFID chip
Unknown
Other...
Patient identifiable information
Clinical data
Patient care logistics
Payment details
Insurance details
Financial information
Employee's personal information
Confidential research data
Confidential organisational data
Copyrighted data
Data under embargo
Medical pictures or recordings
Medical staff
Financial administration staff
Trainee
Personal assistant
Secretary
Admin support
Management/executive/board
Technicians
Cleaners
IT staff
Restaurant catering staff
Volunteers
Other staff
Unknown staff
Patient
Ex-employee
Researcher
Government/police
Other...
Publication of harmful material
Spreading illegal material
Unknown
Delay access
Damage data or facilities
Destroy data
Data lost or gone missing
Locate personal data
Acquire personal data
Disclose personal data
Copy personal data
Networks & Devices
System
Server
Application
Power infrastructure
Security system
Physical barrier
Camera
Smartcard
Hard disk
Documents
Backup tapes
No motive, unintentional action
Challenge
Status
Thrill
Convenience
Covering up errors
Political gain
Emotional
Financial gain
Revenge
Knowledge
Satisfaction
Leisure
Private transport
In a public place
Public transport
Unspecified internal location
Ward
Cafeteria
Corridor
Consultation room

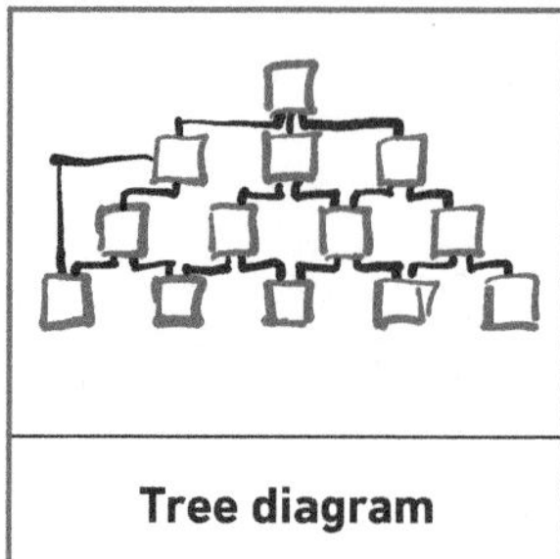

Tree diagram

Software Assurance Maturity Model

OWASP[133]

The Software Assurance Maturity Model (SAMM) is an open framework to help organizations formulate and implement a strategy for software security that is tailored to the specific risks facing the organization. A tree diagram gives an overview of the core business functions of software development with security practices tied to each. The building blocks of the model are the three maturity levels defined for each of the twelve security practices.

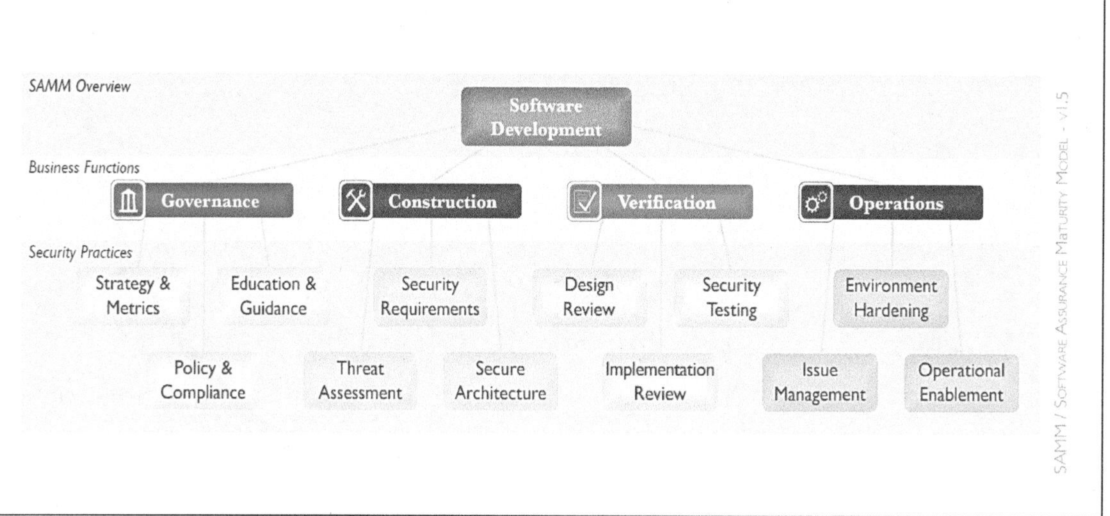

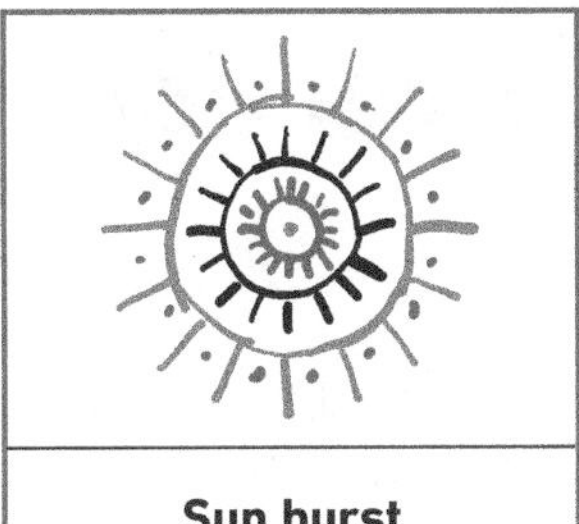

Sun burst

What keeps the CISO busy?
Nicole van Deursen
A sunburst diagram could be an alternative to a bullet-point list job description. You could even play with colours and box sizes to indicate where your set priorities or which tasks take up most of your time.

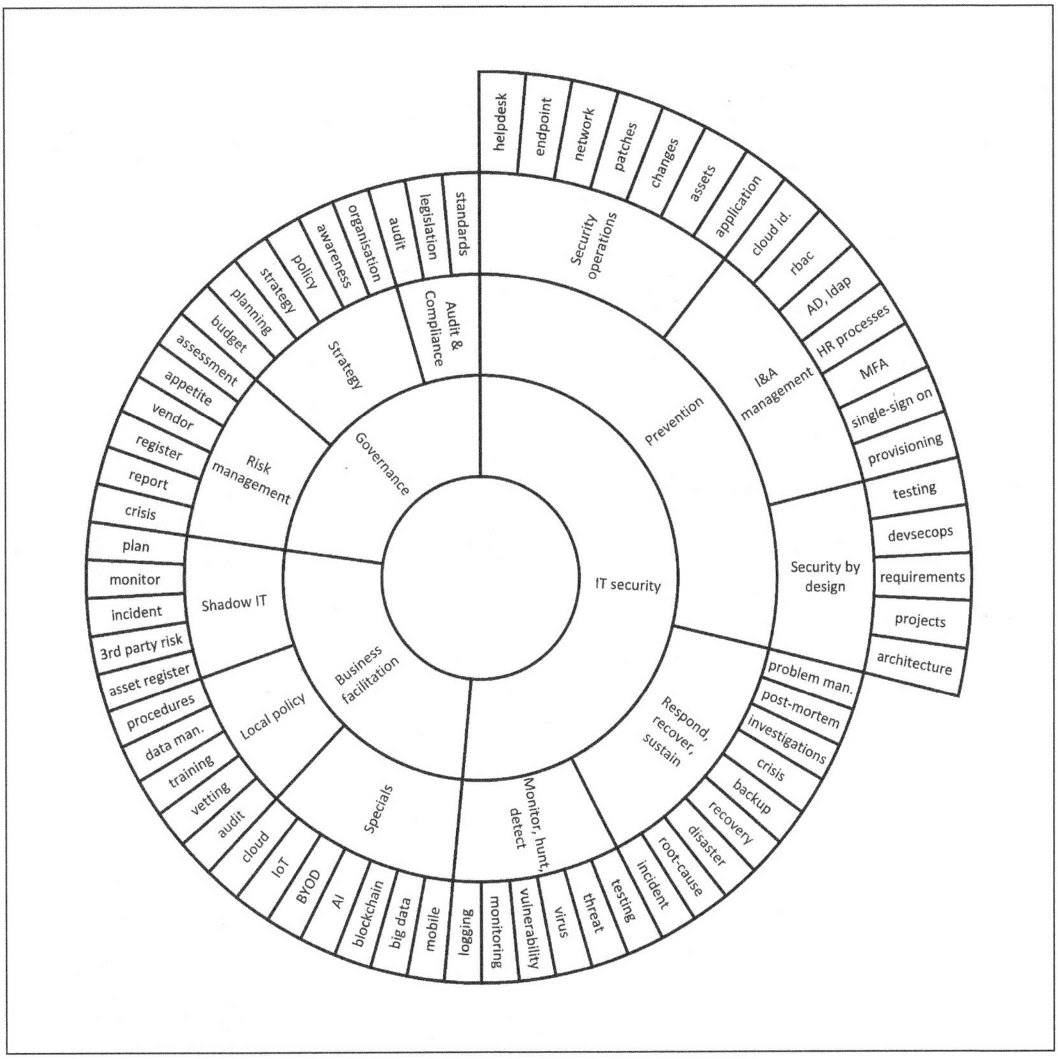

IT security
Governance
Business facilitation
Prevention
Monitor, hunt, detect
Respond, recover, sustain
Strategy
Audit & Compliance
Risk management
Shadow IT
Local policy
Specials
Security operations
I&A management
Security by design
helpdesk
endpoint
network
patches
changes
assets
application
cloud id.
rbac
AD, ldap
HR processes
MFA
single-sign on
provisioning
testing
devsecops
requirements
projects
architecture
problem man.
post-mortem
investigations
crisis
backup
recovery
disaster
root-cause
incident
testing
threat
virus
vulnerability
monitoring
logging
mobile
big data
blockchain
AI
BYOD
IoT
cloud
audit
vetting
training
data man.
procedures
asset register
3rd party risk
incident
monitor
plan
crisis
report
register
vendor
appetite
assessment
budget
planning
strategy
policy
awareness
organisation
audit
legislation
standards

14. Visualizing flow and direction

Cybersecurity work is often very dynamic. Flow and direction diagrams help to visualize how the workflows and what is expected of people. Visualizations help to order a sequence of events or can express a growth vision for the future. The diagrams can be directional, circular, or suggestive of a process. They have in common that they represent movement of some kind.

Flowchart	Bow Tie	Roadmap
Metromap	Comparison	Timeline
Circle	Ladder	Swim lane
Fishbone	Petri net	Maturity model
(Interactive) Pathways	Spiral	Sankey diagram

References to illustrations other than the ones presented in this chapter are:

Petri Net: Petri net modelling results for FTP attacks[134], Petri net model of physical systems and operations for situational awareness[135]
Timeline: The Zeus Timeline (2007-2016)[136]
Sankey: virtual infrastructure[137]
Maturity model: security awareness[138]
Metro map: ransomware families[139]

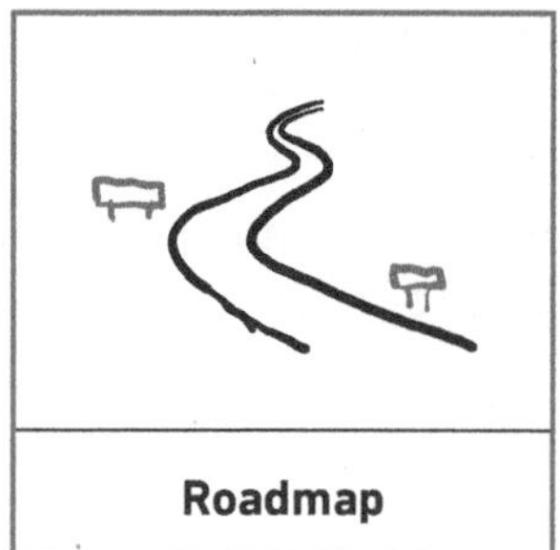

Roadmap

European Cyber Security Month Roadmap
ENISA[140]

European Cyber Security Month (ECSM) is an EU advocacy campaign that takes place in October to promote cybersecurity awareness among citizens. It summarizes the vision for the future: a call-to-action to all organizations to participate in the annual awareness month. On the road to the future are already many cars, each representing one of the partners involved in the awareness month. An earlier chapter in this book discussed the Gestalt principles for message presentation. Which of the principles do you recognize in this picture?

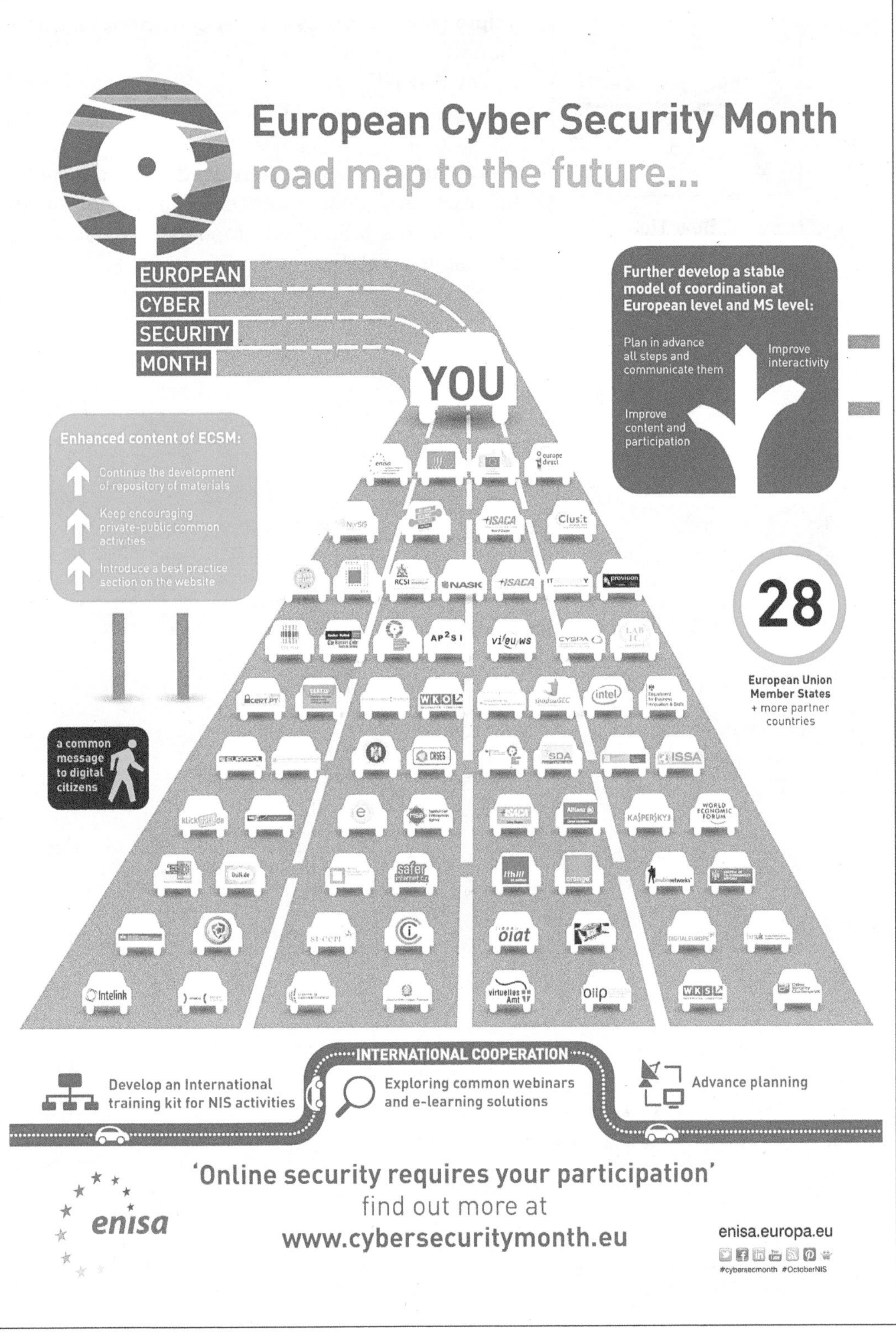
European Cyber Security Month
road map to the future...
EUROPEAN
CYBER
SECURITY
MONTH
YOU
Further develop a stable model of coordination at European level and MS level:
Plan in advance all steps and communicate them
Improve interactivity
Improve content and participation
Enhanced content of ECSM:
Continue the development of repository of materials
Keep encouraging private-public common activities
Introduce a best practice section on the website
a common message to digital citizens
28
European Union Member States
+ more partner countries
INTERNATIONAL COOPERATION
Develop an International training kit for NIS activities
Exploring common webinars and e-learning solutions
Advance planning
'Online security requires your participation'
find out more at
www.cybersecuritymonth.eu
enisa
enisa.europa.eu
#cybersecmonth #OctoberNIS

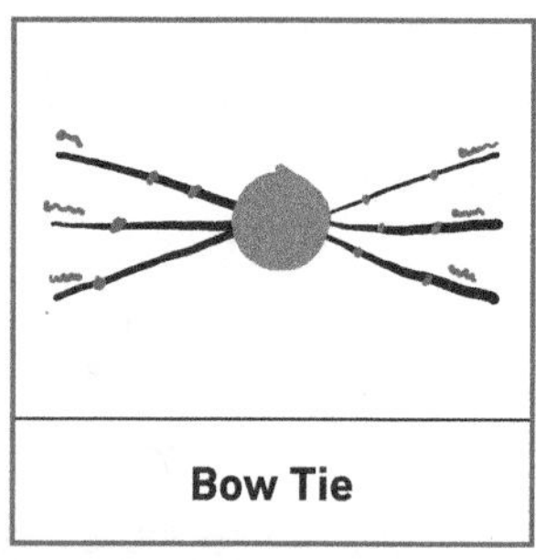

Bow Tie

Triple Bow Tie 5G Cyber Security Supervision Model

Farley Wazir[141]

Copyright:Permission from author

A bow tie analysis is often used to model threats, hazards, and consequences. The author investigated for his MSc thesis models of trust and supervision in relation to the 5G network.

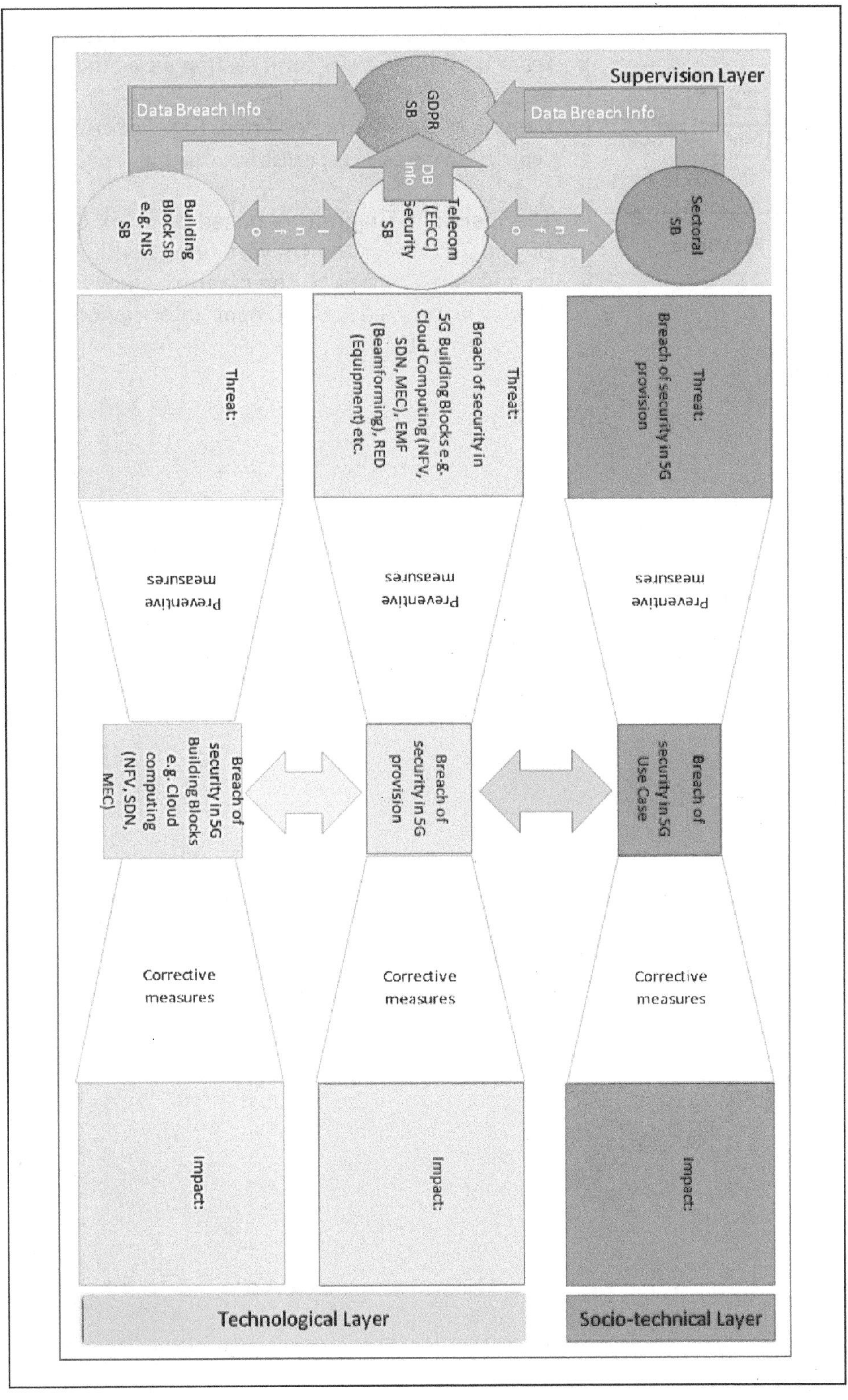
Supervision Layer
Data Breach Info
GDPR SB
Data Breach Info
DB Info
Building Block SB e.g. NIS SB
Info
Telecom (EECC) Security SB
Info
Sectoral SB
Threat:
Threat: Breach of security in 5G Building Blocks e.g. Cloud Computing (NFV, SDN, MEC), EMF (Beamforming), RED (Equipment) etc.
Threat: Breach of security in 5G provision
Preventive measures
Preventive measures
Preventive measures
Breach of security in 5G Building Blocks e.g. Cloud computing (NFV, SDN, MEC)
Breach of security in 5G provision
Breach of security in 5G Use Case
Corrective measures
Corrective measures
Corrective measures
Impact:
Impact:
Impact:
Technological Layer
Socio-technical Layer

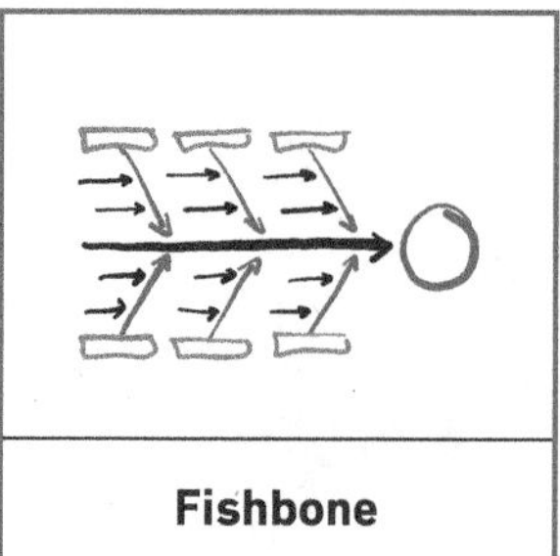

Fishbone

Barriers that slow/hinder/prevent companies from managing their information as a business asset.

Danette McGilvray, James Price, Tom Redman[142]

The Fishbone Diagram is based on work done by Dr Nina Evans of the University of South Australia and James Price[143]. The diagram shows at high level the root causes of poor information asset management.

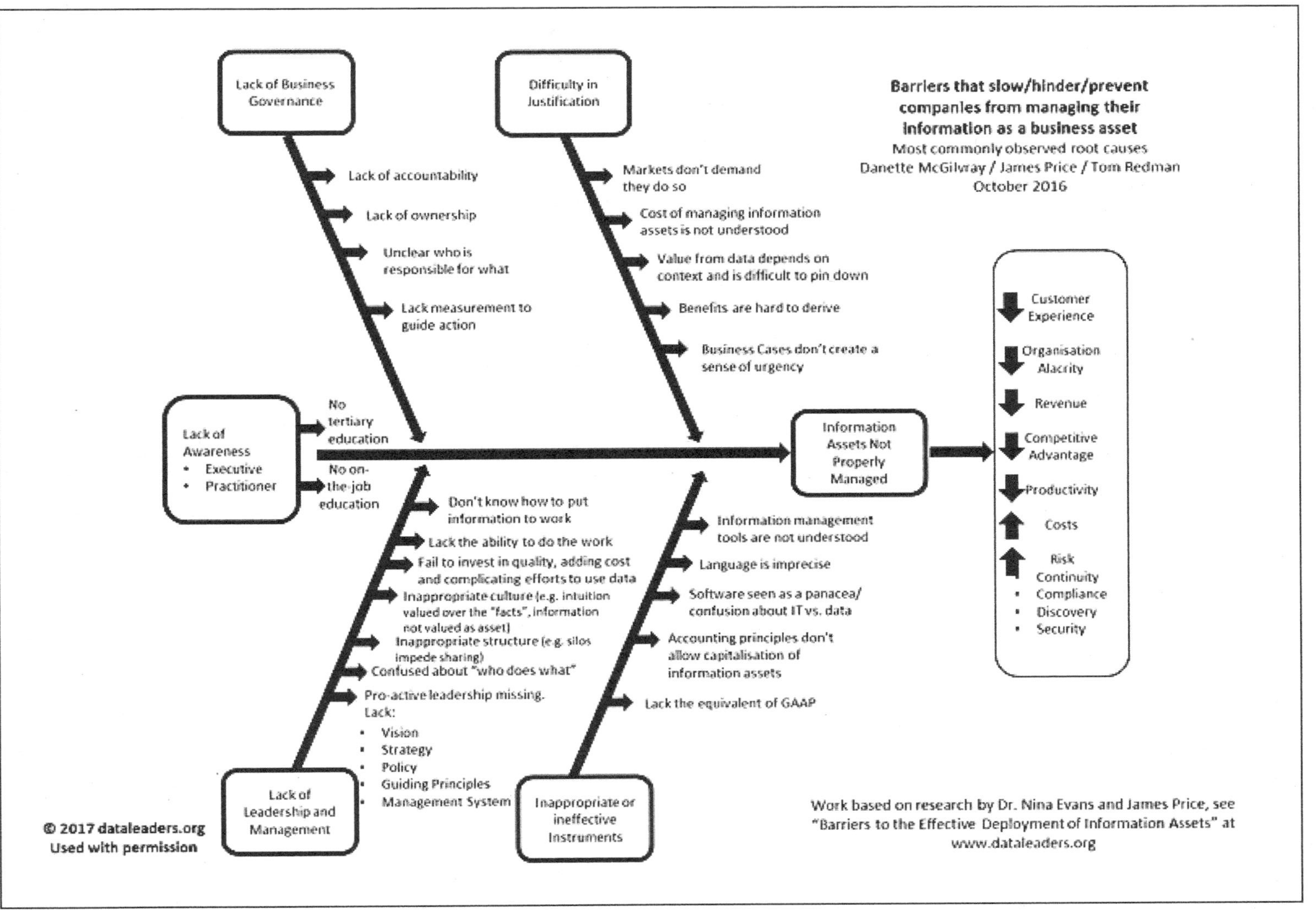

Barriers that slow/hinder/prevent companies from managing their information as a business asset
Most commonly observed root causes
Danette McGilvray / James Price / Tom Redman
October 2016
Lack of Business Governance
Lack of accountability
Lack of ownership
Unclear who is responsible for what
Lack measurement to guide action
Difficulty in Justification
Markets don't demand they do so
Cost of managing information assets is not understood
Value from data depends on context and is difficult to pin down
Benefits are hard to derive
Business Cases don't create a sense of urgency
Lack of Awareness
• Executive
• Practitioner
No tertiary education
No on-the-job education
Lack of Leadership and Management
Don't know how to put information to work
Lack the ability to do the work
Fail to invest in quality, adding cost and complicating efforts to use data
Inappropriate culture (e.g. intuition valued over the "facts", information not valued as asset)
Inappropriate structure (e.g. silos impede sharing)
Confused about "who does what"
Pro-active leadership missing.
Lack:
• Vision
• Strategy
• Policy
• Guiding Principles
• Management System
Inappropriate or ineffective Instruments
Information management tools are not understood
Language is imprecise
Software seen as a panacea/ confusion about IT vs. data
Accounting principles don't allow capitalisation of information assets
Lack the equivalent of GAAP
Information Assets Not Properly Managed
Customer Experience
Organisation Alacrity
Revenue
Competitive Advantage
Productivity
Costs
Risk
Continuity
• Compliance
• Discovery
• Security
© 2017 dataleaders.org
Used with permission
Work based on research by Dr. Nina Evans and James Price, see "Barriers to the Effective Deployment of Information Assets" at www.dataleaders.org

Timeline

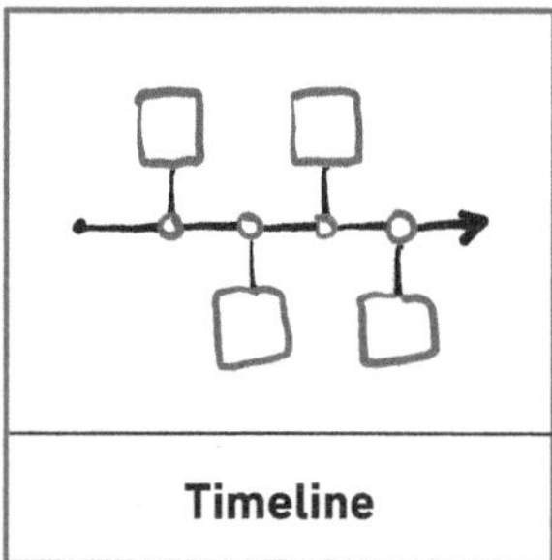

50 years of cybersecurity taxonomies
Nicole van Deursen, created with Timegraphics.
Based on: [25]

This timeline is an extended version of the overview of taxonomies in[25] and lists a selection of academic, public sector, and commercial taxonomies that were published over time.

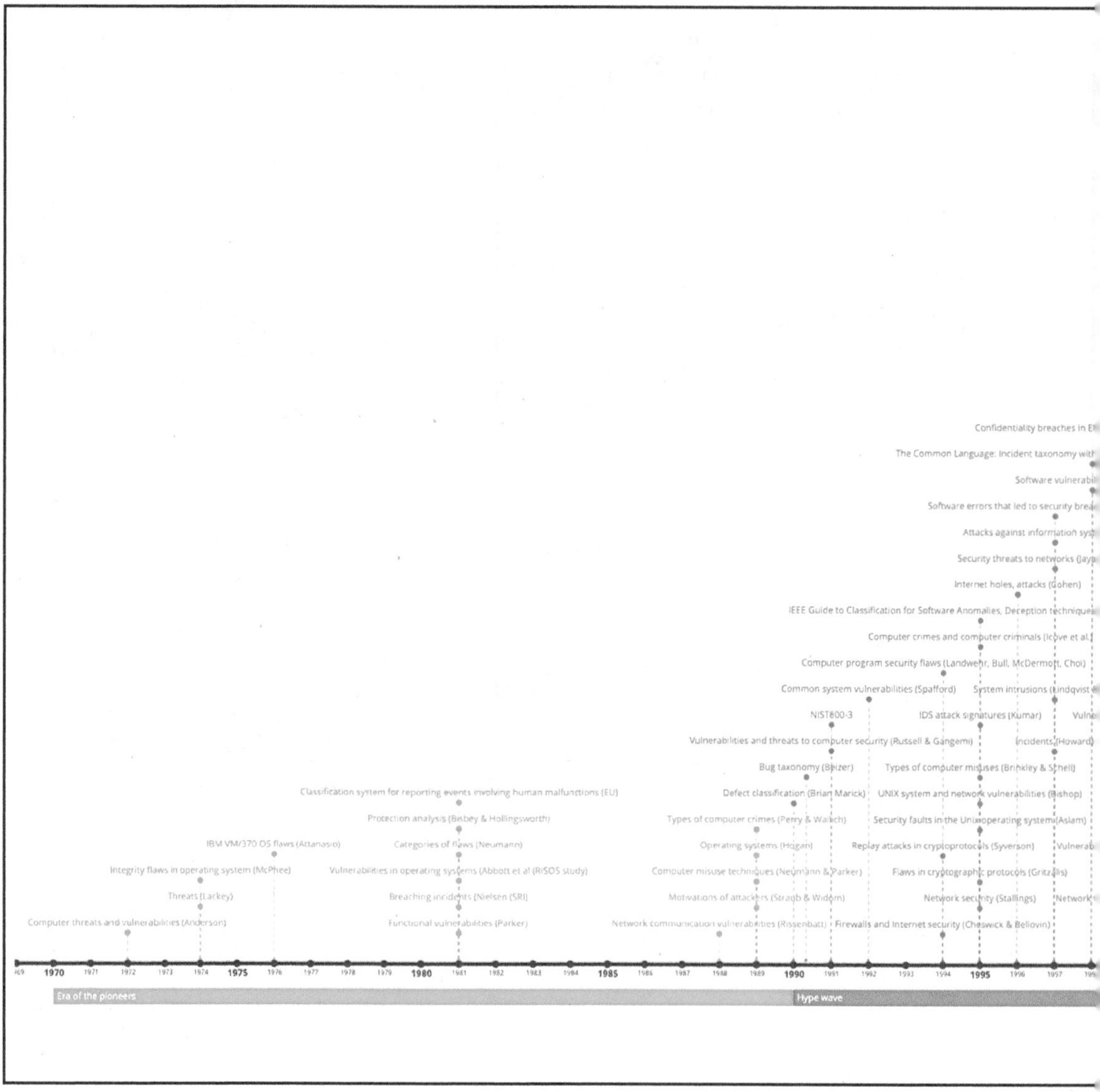

Confidentiality breaches in
The Common Language: Incident taxonomy
Software
Software errors that led to security
Attacks against information
Security threats to networks
Internet holes, attacks (Cohen)
IEEE Guide to Classification for Software Anomalies, Deception
Computer crimes and computer criminals (Icove et al.)
Computer program security flaws (Landwehr, Bull, McDermott, Choi)
Common system vulnerabilities (Spafford)
System intrusions
NIST800-3
IDS attack signatures (Kumar)
Vulnerabilities and threats to computer security (Russell & Gangemi)
Incidents (Howard)
Bug taxonomy (Beizer)
Types of computer misuses (Brinkley & Schell)
Classification system for reporting events involving human malfunctions (EU)
Defect classification (Brian Marick)
UNIX system and network vulnerabilities (Bishop)
Protection analysis (Bisbey & Hollingsworth)
Types of computer crimes (Perry & Wallich)
Security faults in the Unix operating system (Aslam)
IBM VM/370 OS flaws (Attanasio)
Categories of flaws (Neumann)
Operating systems (Hogan)
Replay attacks in cryptoprotocols (Syverson)
Integrity flaws in operating system (McPhee)
Vulnerabilities in operating systems (Abbott et al (RISOS study)
Computer misuse techniques (Neumann & Parker)
Flaws in cryptographic protocols (Gritzalis)
Threats (Lackey)
Breaching incidents (Nielsen (SRI)
Motivations of attackers (Straub & Widom)
Network security (Stallings)
Computer threats and vulnerabilities (Anderson)
Functional vulnerabilities (Parker)
Network communication vulnerabilities (Rissenbatt)
Firewalls and Internet security (Cheswick & Bellovin)
1970
1971
1972
1973
1974
1975
1976
1977
1978
1979
1980
1981
1982
1983
1984
1985
1986
1987
1988
1989
1990
1991
1992
1993
1994
1995
1996
1997
Era of the pioneers
Hype wave

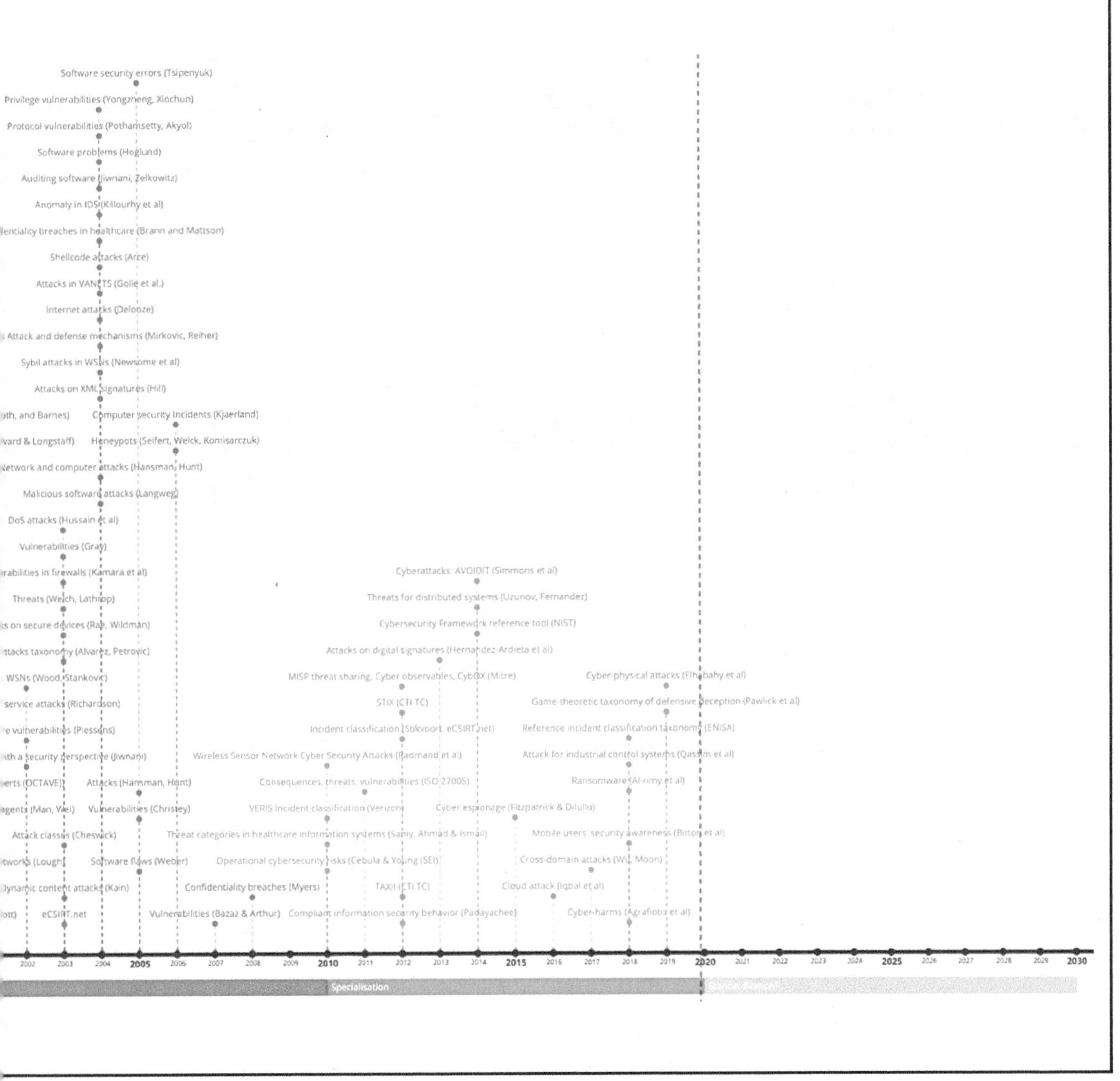

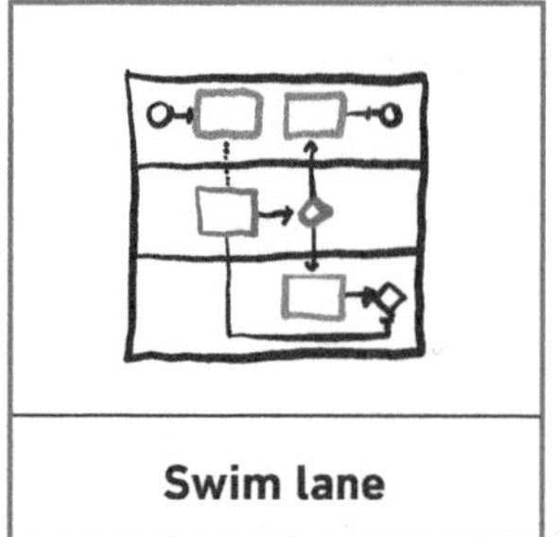

Swim lane

Infection chain of Bad Rabbit Ransomware

Trend Micro[144]

This process flow uses symbols, numbers, text, and arrows to demonstrate the steps that lead up to a ransomware infection. The Gestalt principle of common regions (elements that are part of a group are located within the same closed region) supports the reader by grouping steps that belong together within a box. The arrows support the understanding of direction and sequence in which to read it.

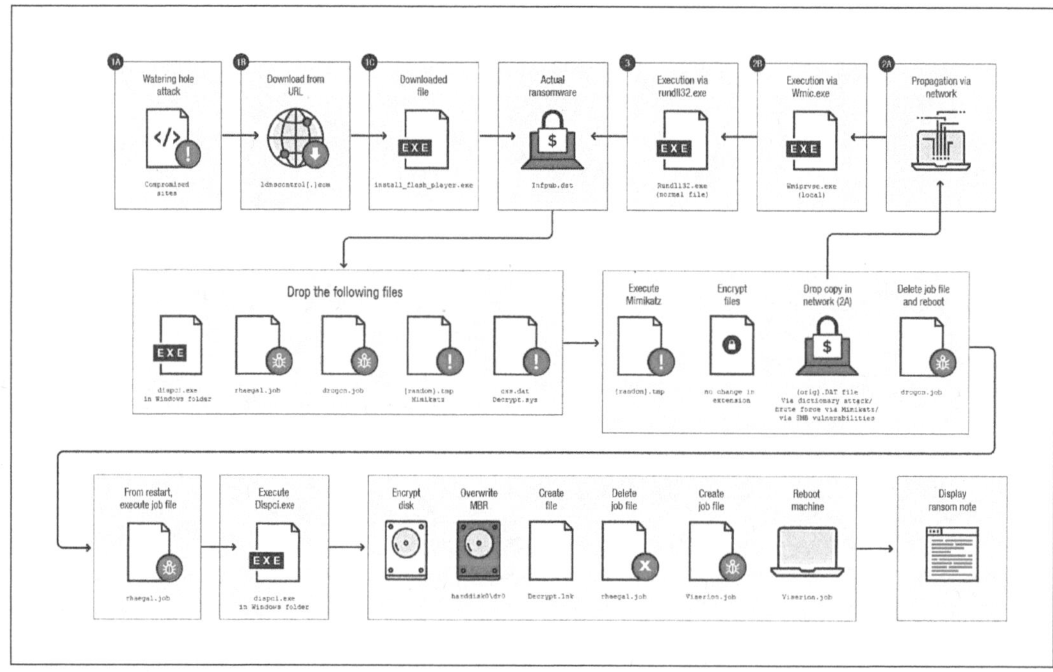

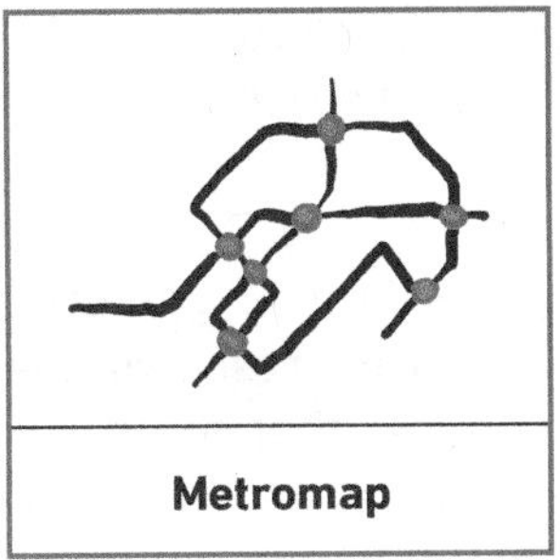

Metromap

Subway map to agile practices

Agile Alliance[145]

Printed with permission from Agile Alliance

The subway map is a screen print from an interactive subway map that shows various aspects of Scrum, Lean, Extreme Programming and others. Each stop on the subway line shows a term associated with the methods.

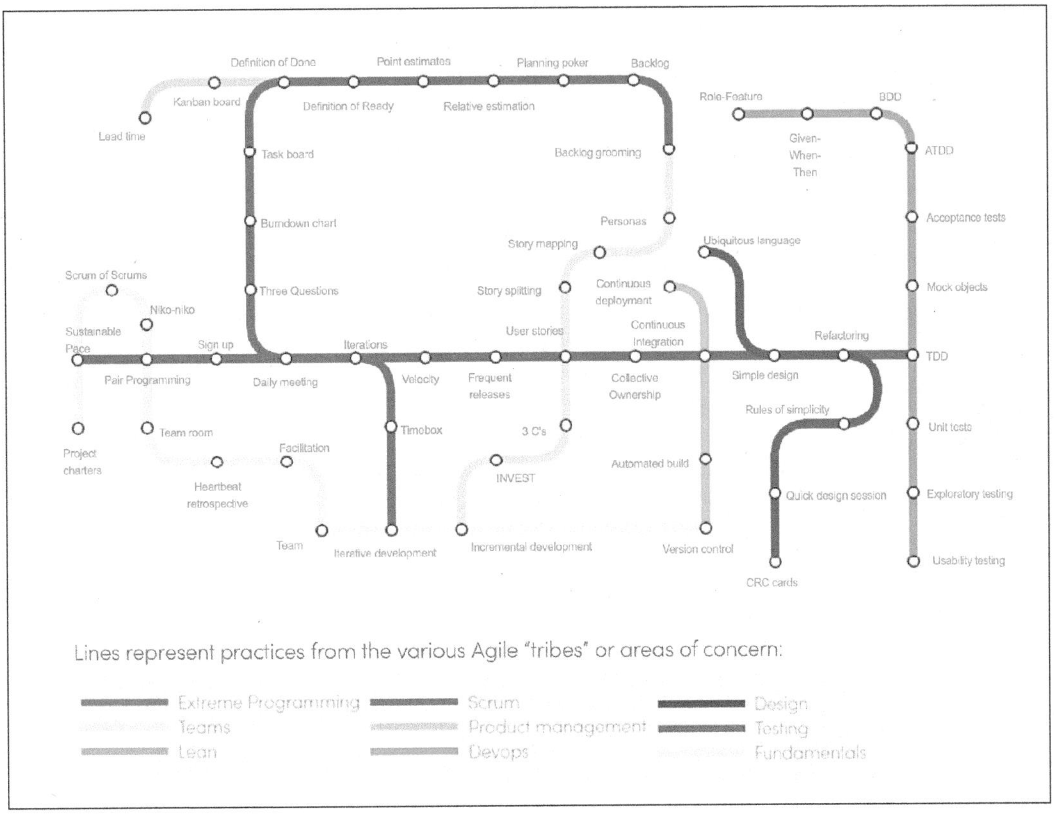

Sankey diagram

Information Risk Scenarios in Healthcare
Nicole van Deursen, created with Flourish
Based on: [25]

This diagram is based on the dataset gathered from organisations in healthcare. It shows the variables and their sub-categories that contributed to information security incidents that were registered between 2006 and 2010.

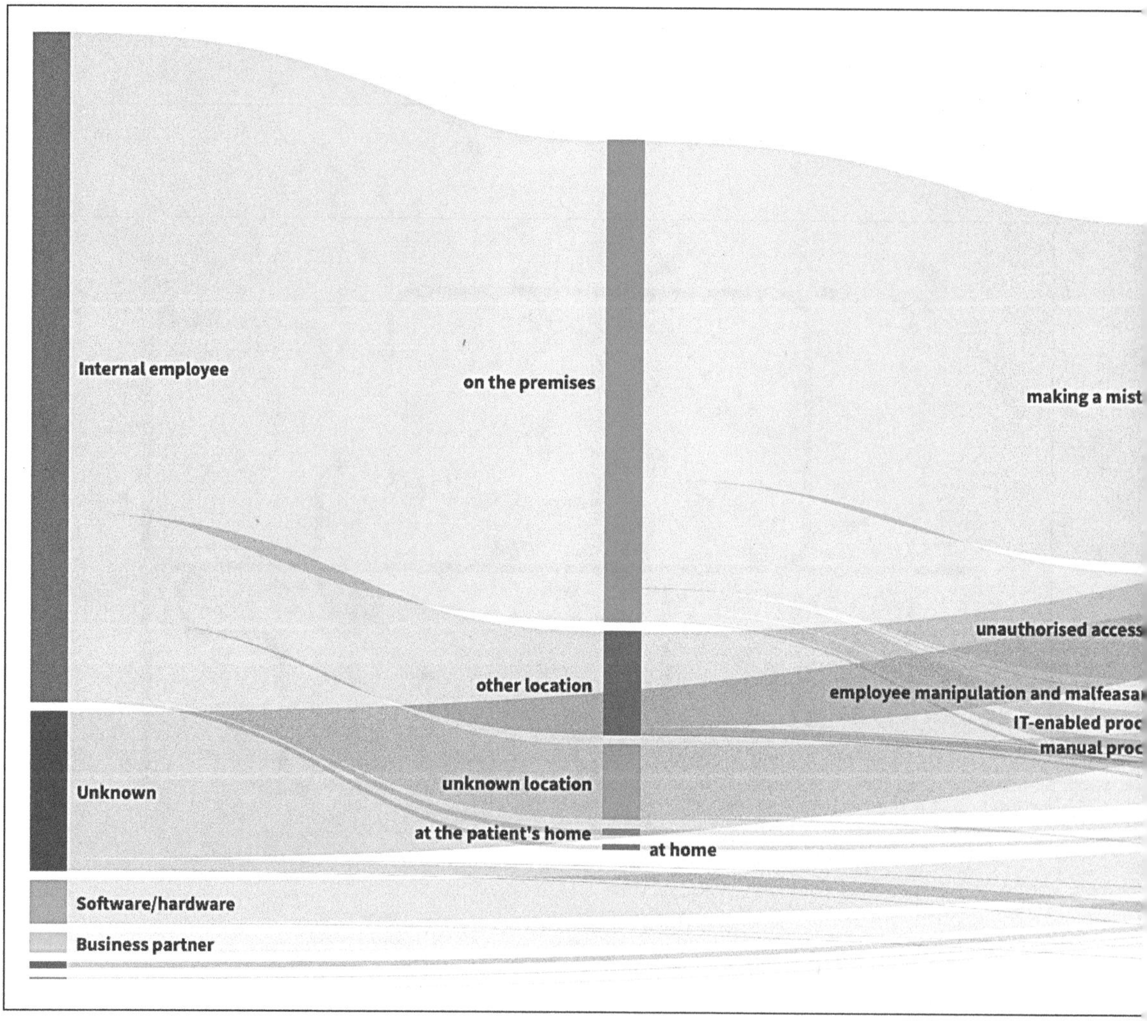

Internal employee
Unknown
Software/hardware
Business partner
on the premises
other location
unknown location
at the patient's home
at home
making a mist
unauthorised access
employee manipulation and malfeasa
IT-enabled proc
manual proc

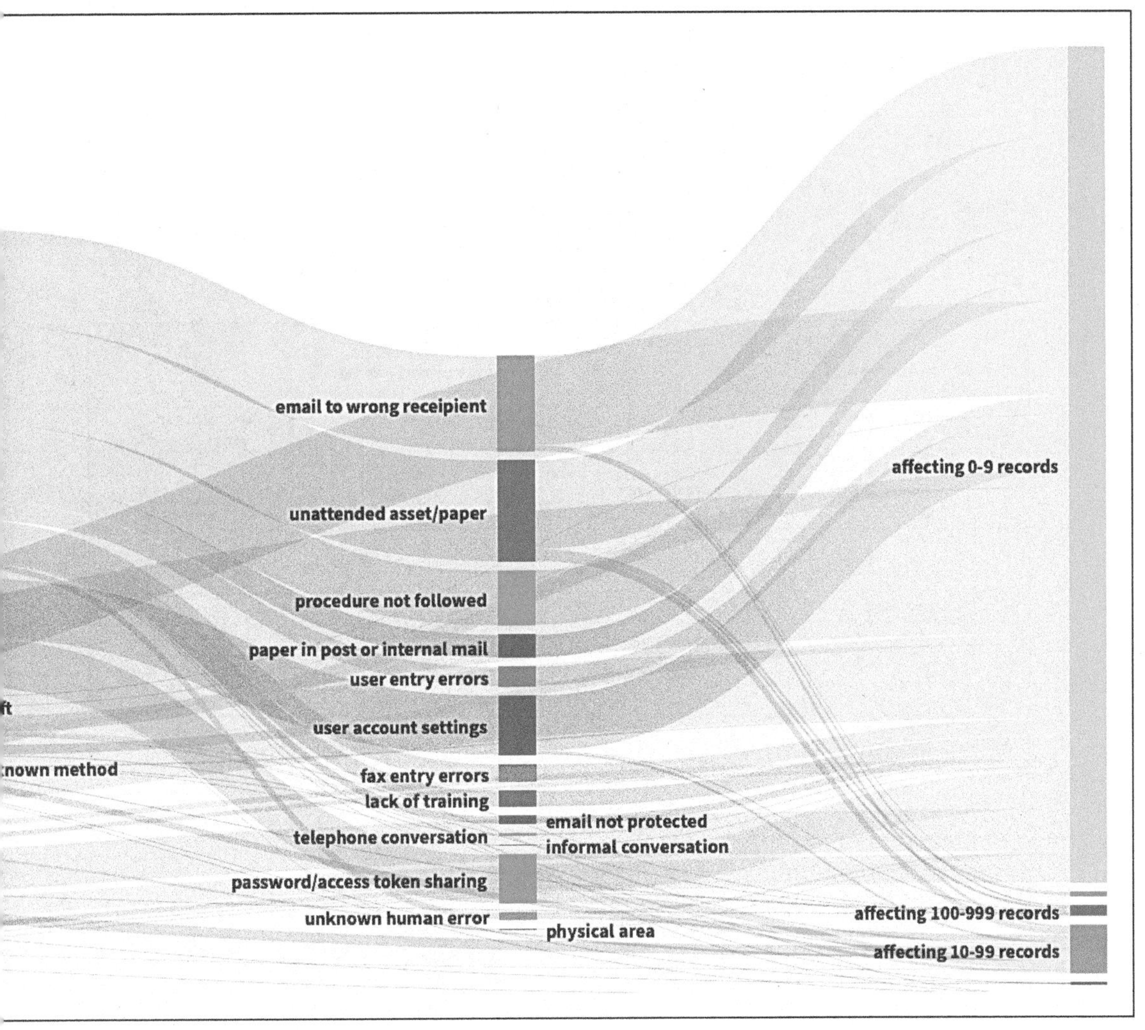
email to wrong receipient
unattended asset/paper
procedure not followed
paper in post or internal mail
user entry errors
user account settings
fax entry errors
lack of training
telephone conversation
password/access token sharing
unknown human error
email not protected
informal conversation
physical area
ft
nown method
affecting 0-9 records
affecting 100-999 records
affecting 10-99 records

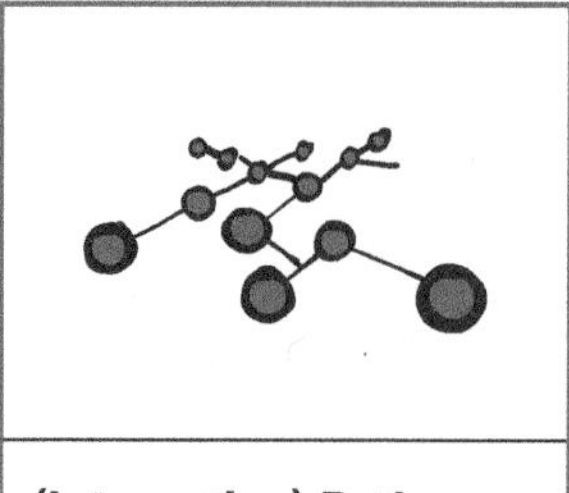

(Interactive) Pathways

CyberSeek

Cybersecurity career pathway[146]

Copyright: permission from CyberSeek Project and partners Burning Glass Technologies, CompTIA, National Initiative for Cybersecurity Education (NICE).

This is an example of an interactive network diagram. It is a screenshot of the CyberSeek website. This interactive career pathway shows key jobs within cybersecurity, common transition opportunities between them, and detailed information about the salaries, credentials, and skill sets associated with each role. The tool supports the concept of a shared language within the community by utilizing the NICE Cybersecurity Workforce Framework to help describe each role as well as listing common job titles that employers use for the same roles.

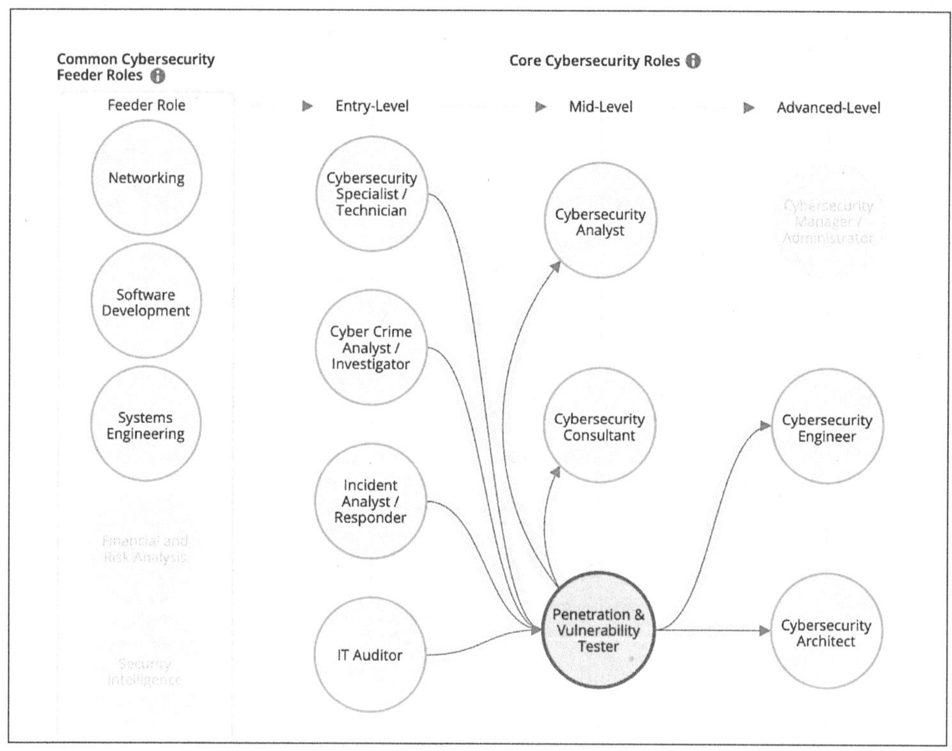

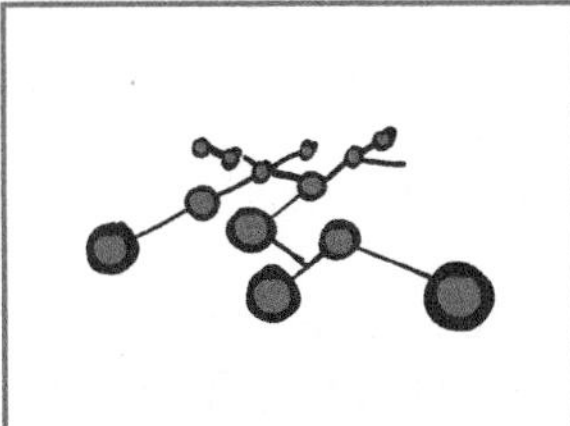

(Interactive) Pathways

The Global Risks Interconnections Map 2018
World Economic Forum, design by Moritz Stefaner / Truth & Beauty[147]
Copyright: permission from WEF and Designer.

This is a screen print of an interactive map that shows how risks are connected. By clicking on the shapes, the map zooms in on the relationships of those risks with other risks. Cyberattacks are visible in purple (purple indicates technological risks). It shows how cyberattacks are connected to many other global risks such as economic risks (in blue) and geopolitical risks (in orange).

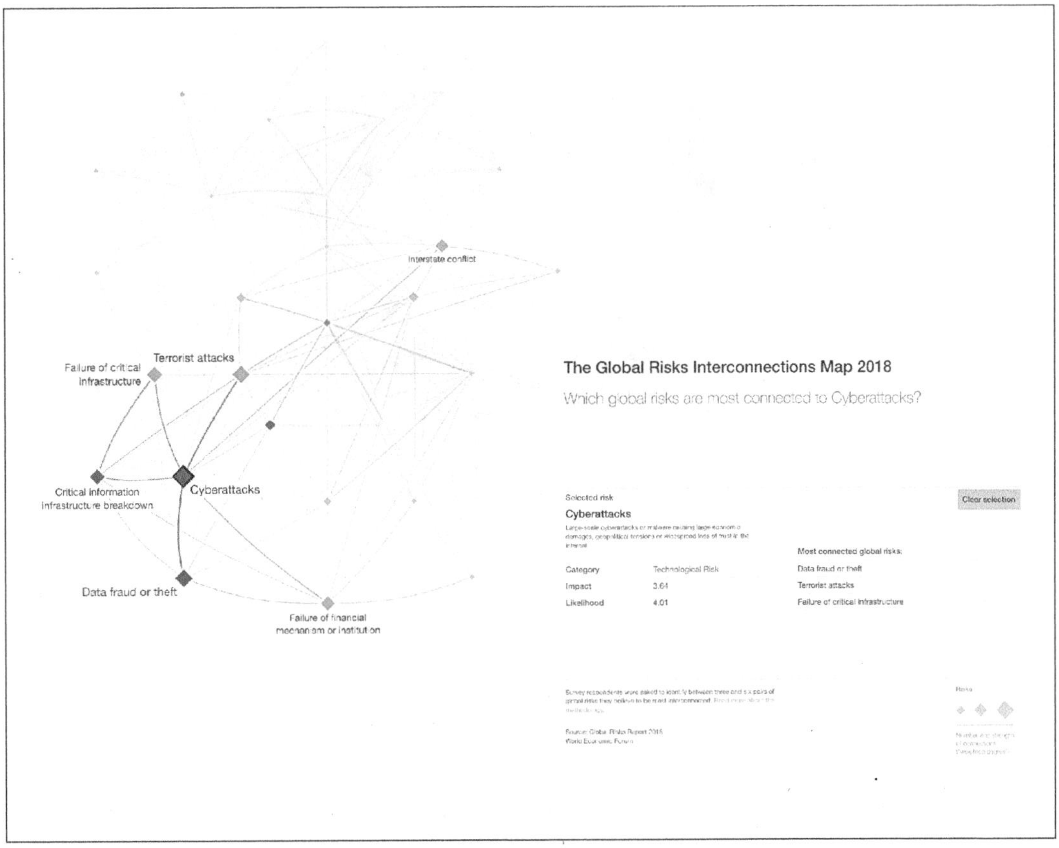

Interstate conflict
Failure of critical infrastructure
Terrorist attacks
Critical information infrastructure breakdown
Cyberattacks
Data fraud or theft
Failure of financial mechanism or institution

The Global Risks Interconnections Map 2018
Which global risks are most connected to Cyberattacks?

Selected risk
Cyberattacks
Clear selection

Category Technological Risk
Impact 3.64
Likelihood 4.01

Most connected global risks:
Data fraud or theft
Terrorist attacks
Failure of critical infrastructure

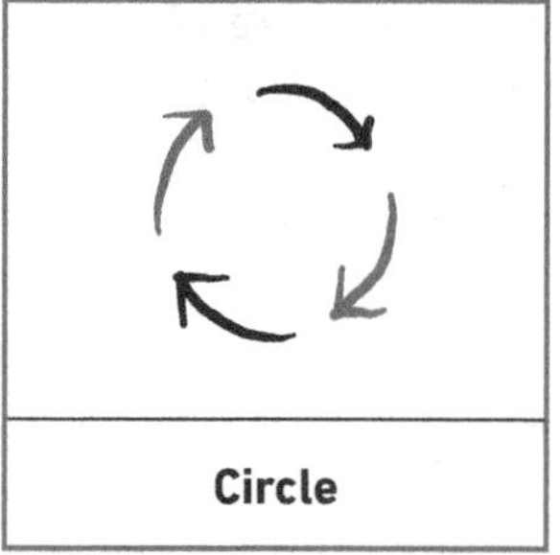

10 steps to cyber security

National Cyber Security Centre, UK[148]

The aim of this infographic is to explain the central focus points of a cybersecurity strategy. The dominant part is the middle figure, a circular process that is continuously moving clockwise.

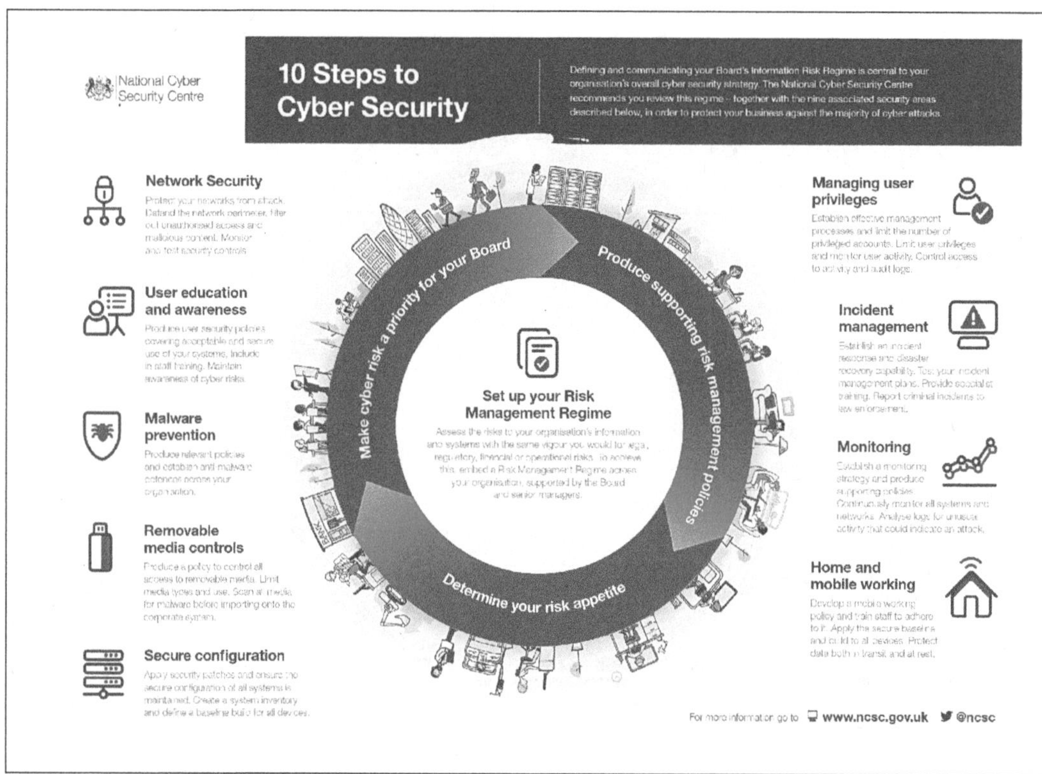

15. Visualizing whole, loose, and connected concepts

In some situations, you want to explain related concepts that do not involve a hierarchy or movement. For example, when things exist in parallel or next to each other. In those situations, you can use simple shapes as circles and rectangles. By making them bigger or smaller, you can express importance, where the shapes cross is an overlap, and how they are connected signifies a relationship.

Bubbles	**Venn diagram**	**Concentric diagram (onion)**
Radar	**Periodic table**	**Geographical map**
A D B C		
Matrix	**Chord chart**	**Sphere**
Parallel coordinates	**Network**	**Arc diagram**

For more examples please refer to:

Bubbles: World's biggest data breaches and hacks[149], network security[150]
Venn: the difference between cybersecurity and information security[151], Venn diagram of the difference between Automotive Cyber security and Automotive Information Security[152]
Periodic table: elements that make a security professional[153]
Network: malware distribution network[154]
Parallel coordinates: anomaly detection[155]
Geographical maps: heatmaps[156] and real-time threat maps[157–161]
Matrix: SWOT analysis of Ransomware as a Service (RaaS) business model[162].

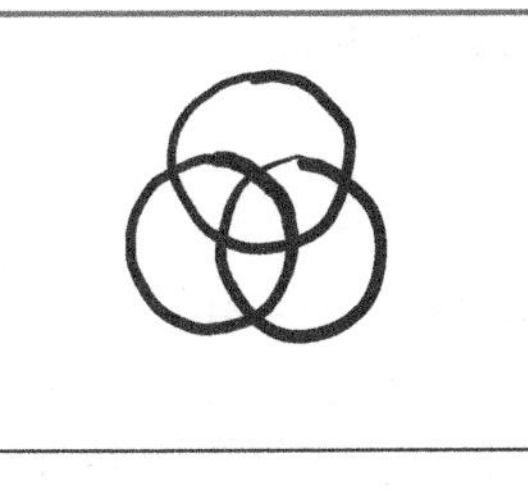

Venn diagram

Data processing activities and compliance obligations.
Pietro Calorio[163]
Permission from author to use the picture.

Drawing a Venn diagram is an easy way to support decision-making when fields of interest or responsibilities overlap. In this case, the author drew a 5-set Venn diagram to describe complex relationships between data processing activities and related tasks to perform.

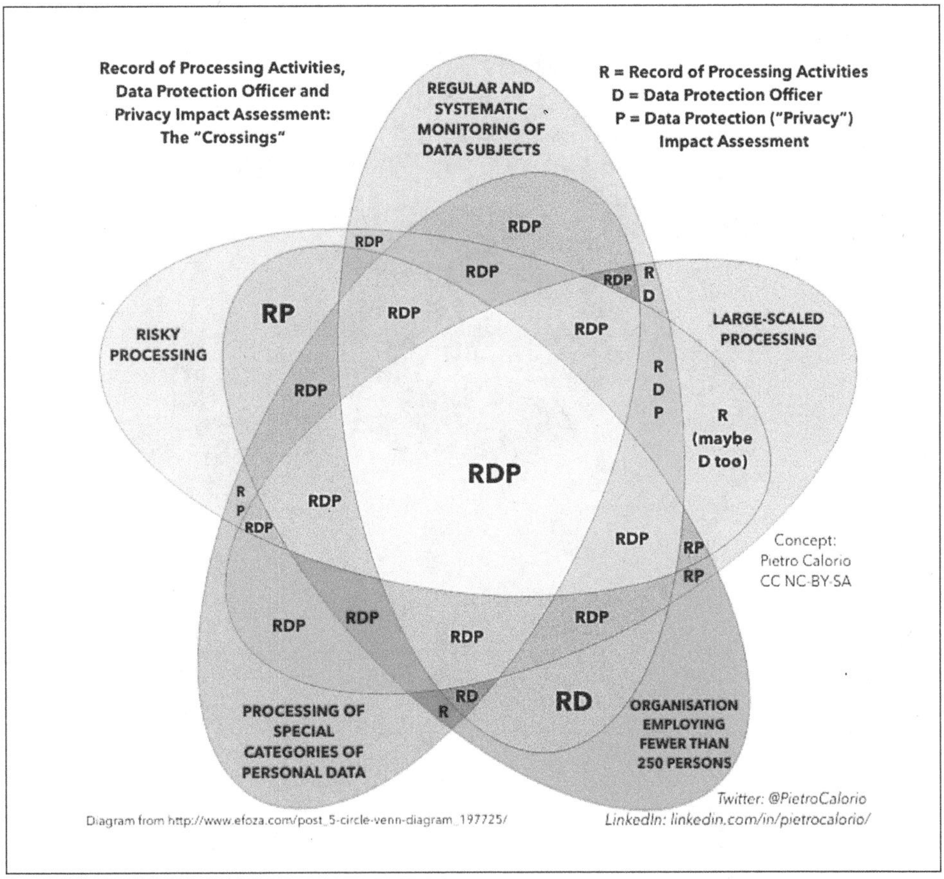

Record of Processing Activities, Data Protection Officer and Privacy Impact Assessment: The "Crossings"

R = Record of Processing Activities
D = Data Protection Officer
P = Data Protection ("Privacy") Impact Assessment

REGULAR AND SYSTEMATIC MONITORING OF DATA SUBJECTS
RISKY PROCESSING
LARGE-SCALED PROCESSING
PROCESSING OF SPECIAL CATEGORIES OF PERSONAL DATA
ORGANISATION EMPLOYING FEWER THAN 250 PERSONS
RP
RDP
RD
R (maybe D too)

Concept: Pietro Calorio CC NC-BY-SA

Diagram from http://www.efoza.com/post_5-circle-venn-diagram_197725/

Twitter: @PietroCalorio
LinkedIn: linkedin.com/in/pietrocalorio/

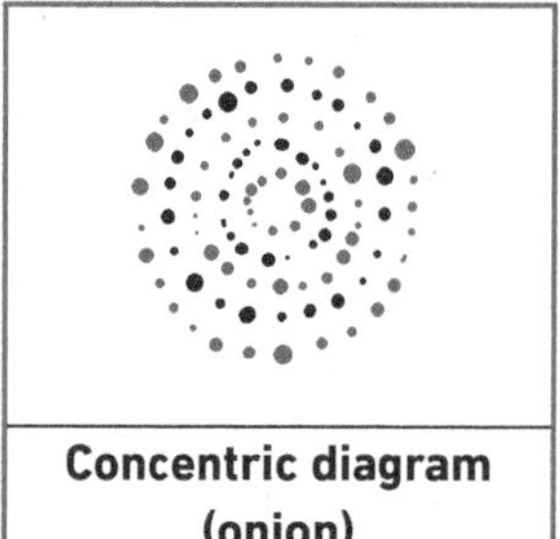

Concentric diagram (onion)

A model for information security leadership
Jeffrey Pomerantz, EDUCAUSE[164]

A working group of information security officers from multiple institutions for higher education collaborated to describe the role of a CISO in higher education. The illustration is a concentric diagram with layers. The innermost ring represents core roles, the next ring represents primary roles, and the third ring discrete roles. The outermost circle is the overarching role of human. Good leaders should not forget traits of compassion, human, and learning from mistakes.

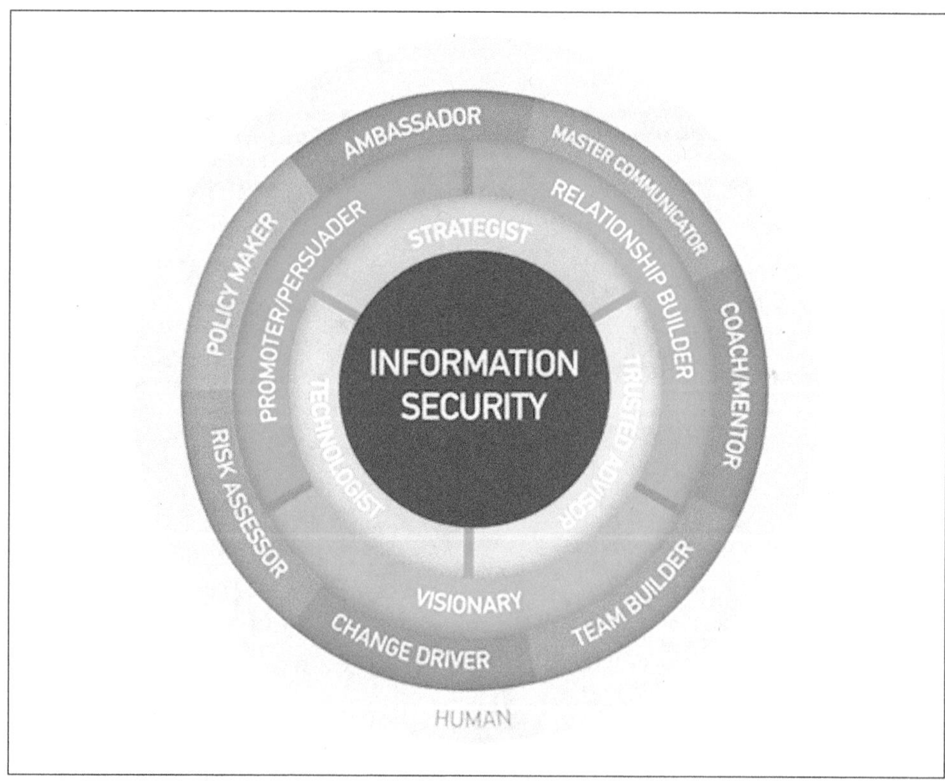

Bubbles

Information security tasks

Nicole van Deursen

Based on: [165]

This chart is based on a content analysis of 52 job vacancies in The Netherlands in spring 2018. In only 52 vacancies, 585 unique tasks were counted. The chart shows tasks grouped into similar areas.

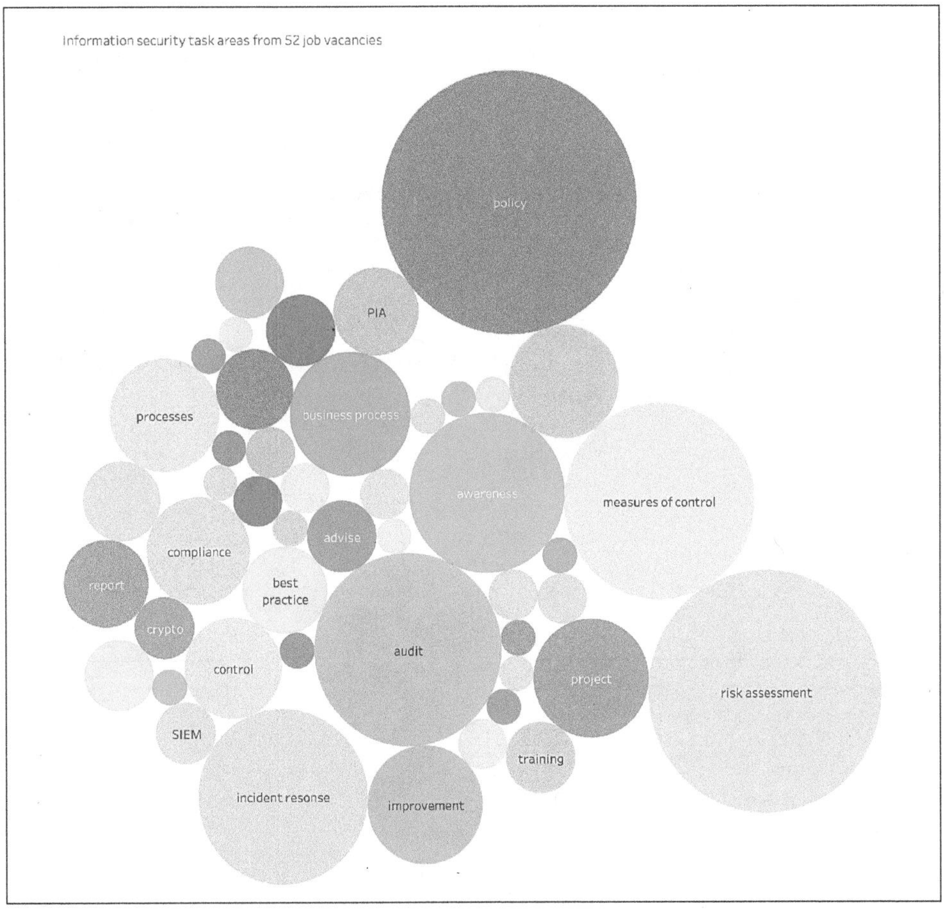

Information security task areas from 52 job vacancies

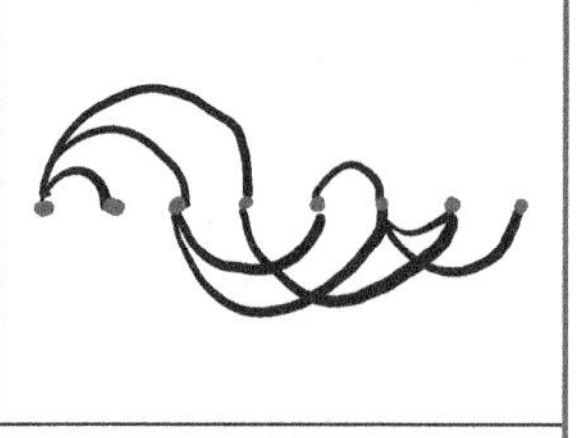

Arc diagram

Visualising attack graphs

Concept and design: LUST (Jeroen Barendse with Eric Li) for TREsPASS Project.

Data: Verizon annual Data Breach Investigations Report (DBIR) 2016[166]

LINKS TO ACTIONS

INCIDENT COUNT

LINKS TO ATTRIBUTES

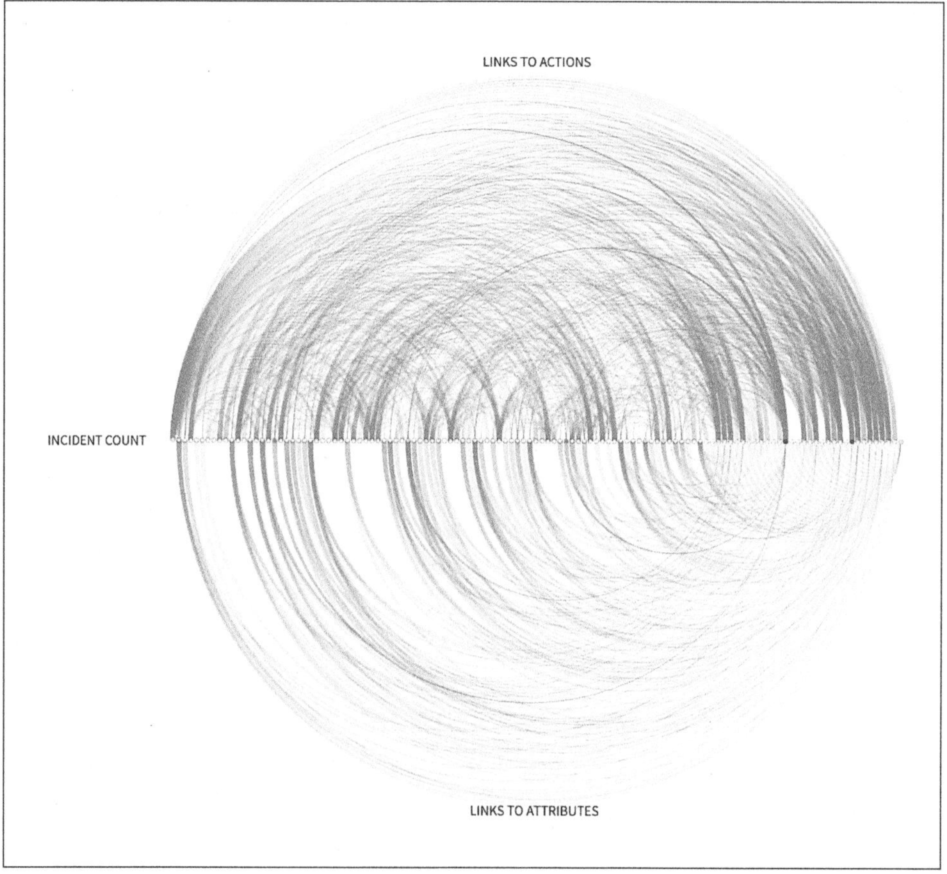

Security researchers can use attack graphs to organise information on possible attack paths within a certain space. The graphs above are based on Verizon's annual Data Breach Investigations Report (2016). They have a starting point at the left and an end point (the goal), as well as nodes that function as attack steps or entities. The edges are possible paths from one entity to another. These nodes and edges carry with them several parameters, such as probability, cost, and incident count.

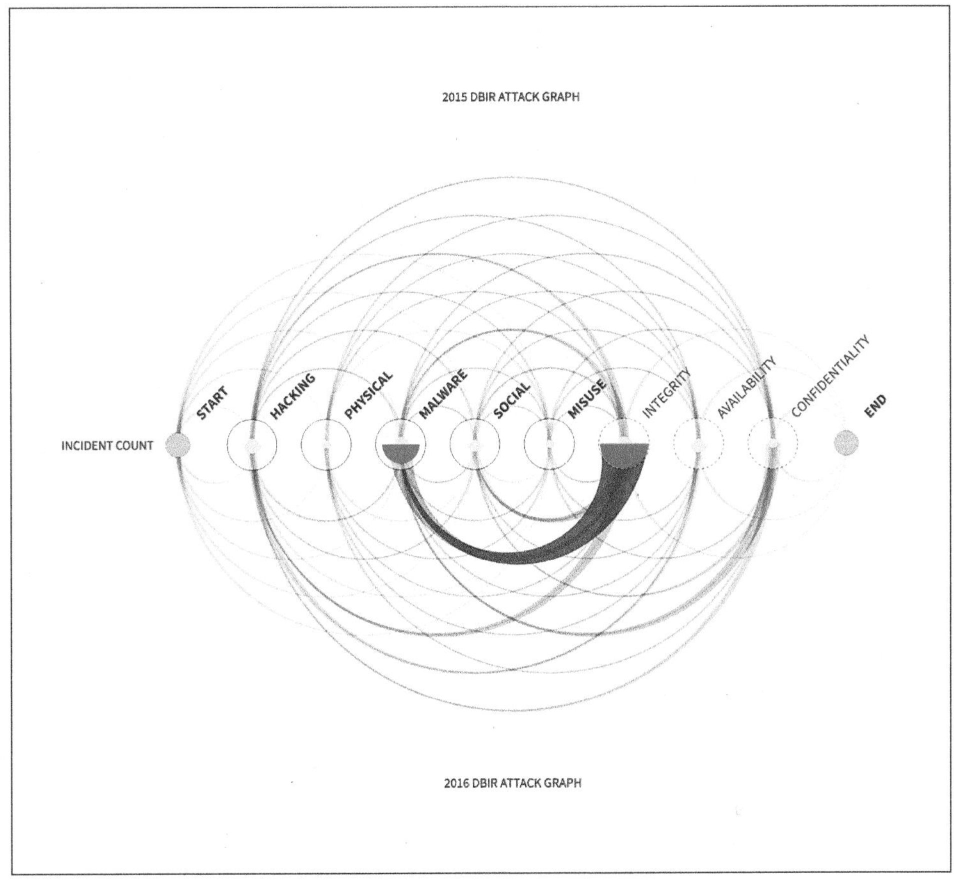

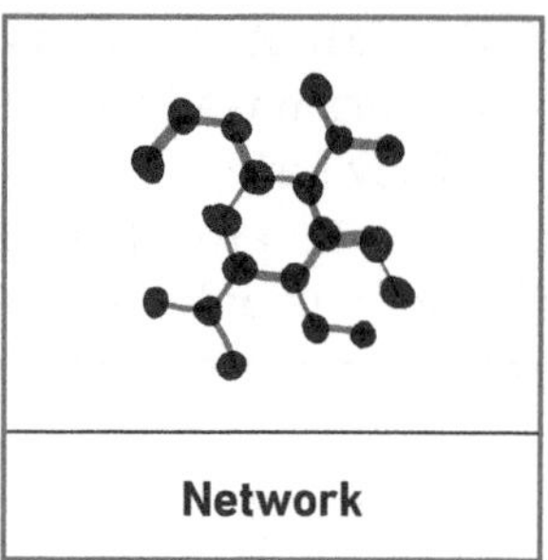

Cybersecurity horoscope 2019
Nicole van Deursen
Based on [167]
This star map summarizes the cybersecurity trend predictions for 2019 from various cybersecurity experts and organisations.

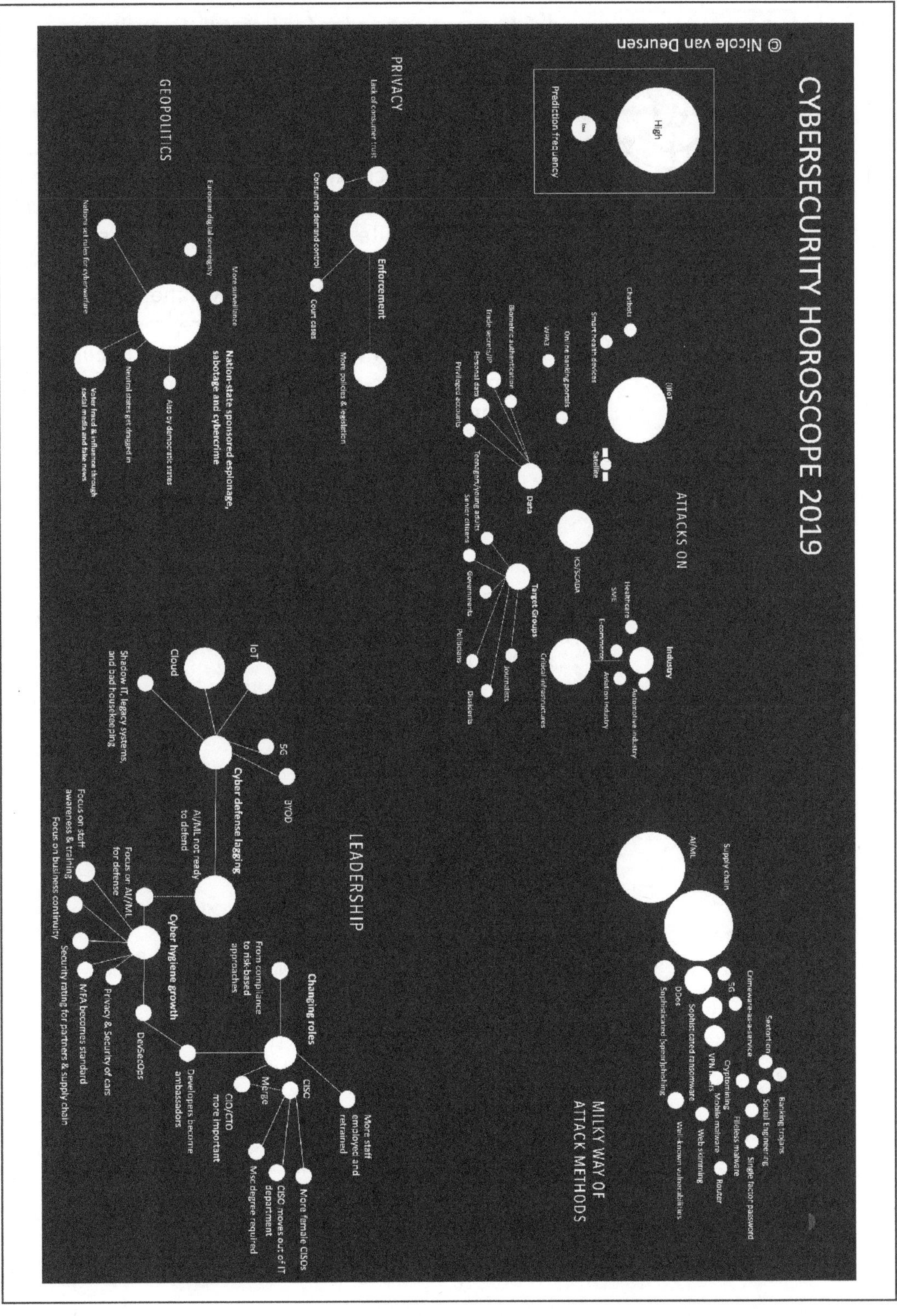
CYBERSECURITY HOROSCOPE 2019
© Nicole van Deursen
Prediction frequency
High
ATTACKS ON
MILKY WAY OF ATTACK METHODS
LEADERSHIP
PRIVACY
GEOPOLITICS
IIoT
Data
Target Groups
Industry
ICS/SCADA
Critical infrastructures
Supply chain
AI/ML
Changing roles
Cyber defense lagging
Cyber hygiene growth
Enforcement
Nation-state sponsored espionage, sabotage and cybercrime
IoT
Cloud
5G
BYOD
DevSecOps
Privacy & Security of cars
MFA becomes standard
Security rating for partners & supply chain
Focus on business continuity
Focus on staff awareness & training
Focus on AI/ML for defense
AI/ML not ready to defend
Shadow IT, legacy systems, and bad housekeeping
From compliance to risk-based approaches
More staff employed and retained
More female CISOs
CISO moves out of IT department
Msc degree required
CIO/CTO more important
Developers become ambassadors
Consumers demand control
Court cases
More policies & legislation
Lack of consumer trust
Voter fraud & influence through social media and fake news

Occupations radar for Safety& Security.
Security Talent, The Hague Security Delta (HSD)[168]
Printed with permission from The Hague Security Delta

The picture of the radar shows occupations in Safety & Security and how they relate to each other. Professions that are close to each other in terms of skills and knowledge or those that often cooperate in practice are placed close to each other on the radar. This follows the Gestalt principle of proximity. The colours represent the clusters based on what type of Safety & Security they aim for. Other similarities are represented by the shape of instances on the radar.

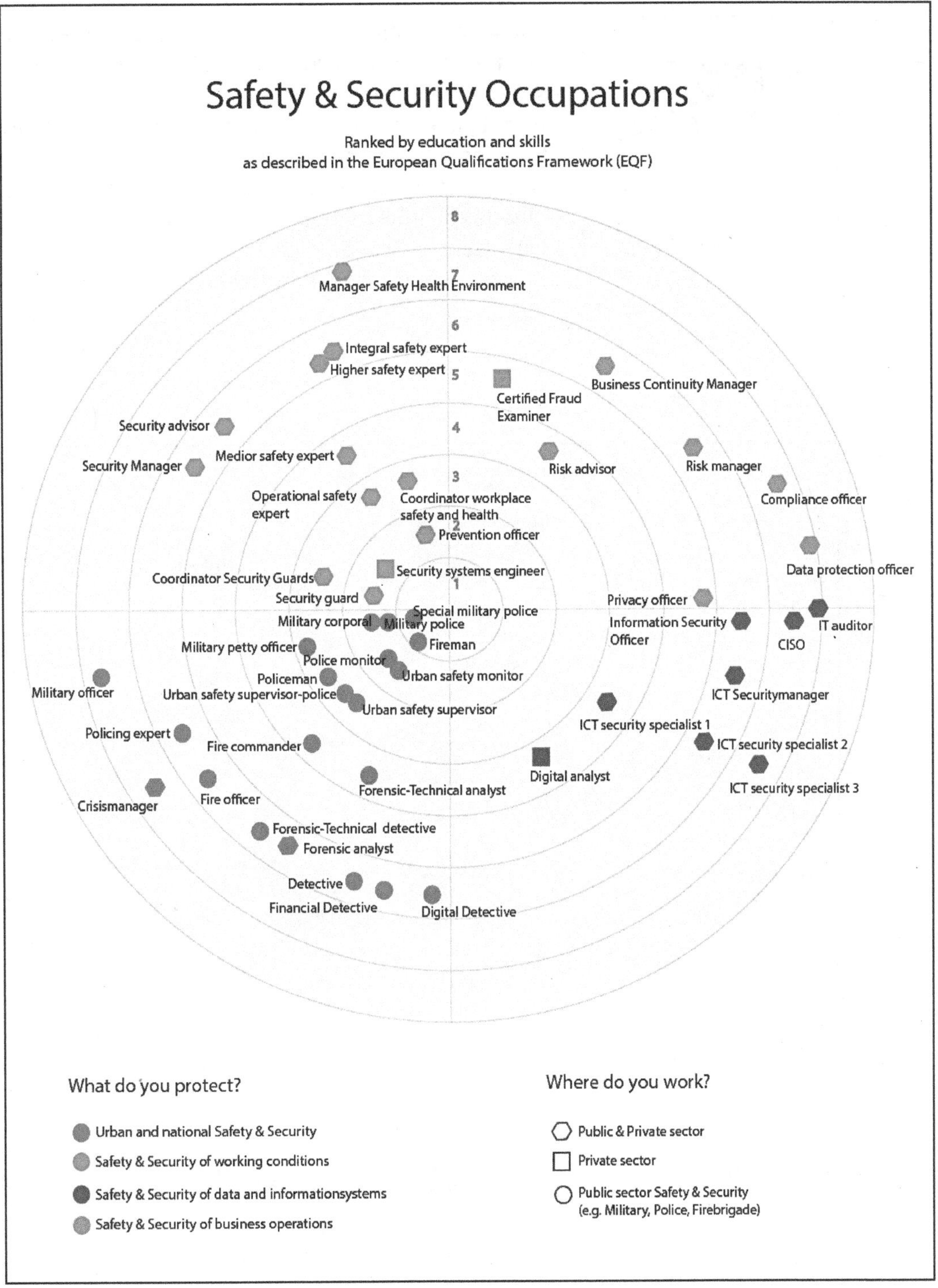
Safety & Security Occupations
Ranked by education and skills
as described in the European Qualifications Framework (EQF)
8
7
6
5
4
3
2
1
Manager Safety Health Environment
Integral safety expert
Higher safety expert
Certified Fraud Examiner
Business Continuity Manager
Security advisor
Medior safety expert
Security Manager
Risk advisor
Risk manager
Compliance officer
Operational safety expert
Coordinator workplace safety and health
Prevention officer
Data protection officer
Coordinator Security Guards
Security systems engineer
Security guard
Privacy officer
Special military police
Military corporal
Military police
Information Security Officer
IT auditor
CISO
Fireman
Military petty officer
Police monitor
Policeman
Urban safety monitor
Military officer
Urban safety supervisor-police
Urban safety supervisor
ICT Securitymanager
ICT security specialist 1
ICT security specialist 2
ICT security specialist 3
Policing expert
Fire commander
Digital analyst
Forensic-Technical analyst
Crisismanager
Fire officer
Forensic-Technical detective
Forensic analyst
Detective
Financial Detective
Digital Detective
What do you protect?
Urban and national Safety & Security
Safety & Security of working conditions
Safety & Security of data and informationsystems
Safety & Security of business operations
Where do you work?
Public & Private sector
Private sector
Public sector Safety & Security
(e.g. Military, Police, Firebrigade)

Geographical map

These maps show the frequency of Google searchers for the words cyber security and information security in different parts of the world and in different years. The intensity of the colour represents the frequency of searches. The red colour shows where the cyber security is mostly searched, and blue is for information security. All maps created with Google Trends. Data source: Google Trends (https://www.google.com/trends).

Searches in 2015	 ● information security ● cyber security
Searches in 2017	 ● information security ● cyber security
Searches in 2019	 ● information security ● cyber security

16. Visual activities

Cybersecurity professionals work in teams on many occasions. This section shows some examples of visual work forms that teams can use in meetings to solve problems, to generate new ideas, or to design new processes or systems. Visual collaboration techniques are helpful during such occasions. There are countless techniques for many different situations and for different audiences.

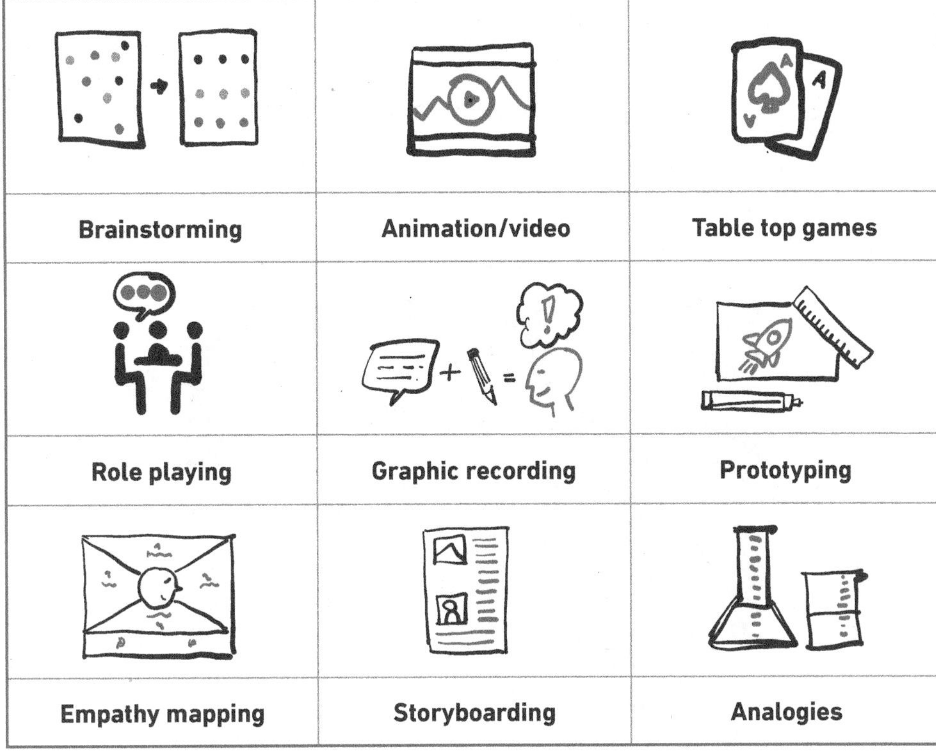

References to illustrations other than the ones presented in this chapter are:

Storyboard: Cyber response[169], Mobile security iOS app[170]
Brainstorming/protoyping: GDPR icons[171]
Video/animations: awareness and training videos[172–175]

Table top games

Cyber Threat Defender Collectible Card Game

The University of Texas at San Antonio's (UTSA) Center for Infrastructure Assurance and Security [176]

Pictures with permission from The University of Texas at San Antonio's (UTSA) Center for Infrastructure Assurance and Security

The card game is designed to teach the essentials of cybersecurity information in the classroom and features asset, defence, attack, and event cards. The game combines learning styles to introduce basic cybersecurity concepts to children.

Graphic recording is the practice of documenting key thoughts and ideas in real-time, summarizing what's happening in the room and transforming it into visuals and words. It creates a visual memory of a meeting or a presentation.

Graphic recording

Securing the Future State

SWIFT Innotribe at Sibos 2018[177]

Presenters: Duena Blomstrom, Jane Frankland.

Drawn by Evan Wondolowski, Collective Next

Printed with permission from SWIFT.

Graphic recording of a presentation at SWIFT Innotribe at the Sibos conference in Sidney, 2018.

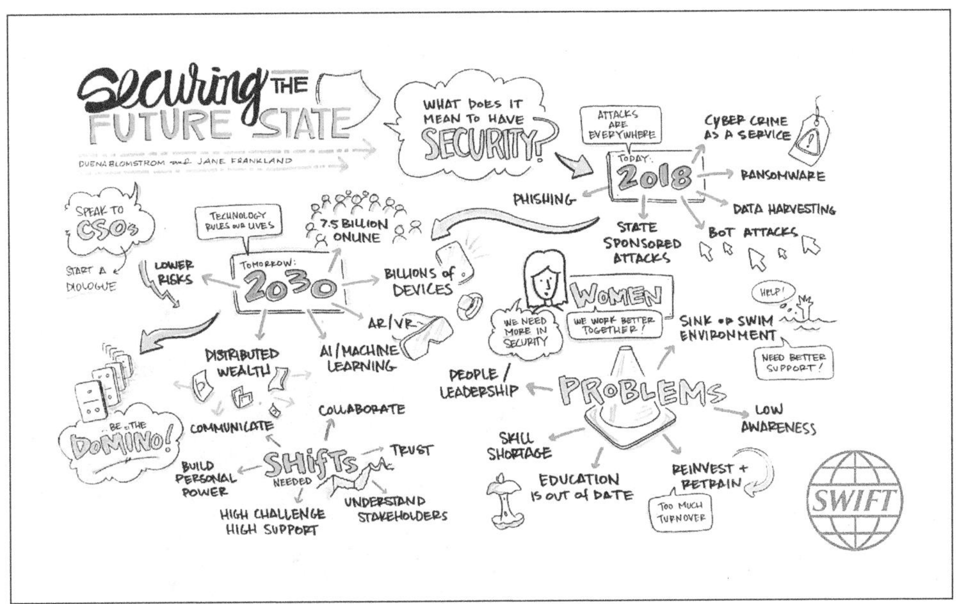

Prototyping

Both pictures: Paper prototyping: Cloud case study
Paper Prototyping method developed by Royal Holloway University with LUST for the Trespass Project[178]

Paper prototyping is a means of creating a paper version of a digital interface and inviting a participant group to engage with the paper prototype simulating the use of the digital interface. This has placed emphasis on taking paper prototypes to user groups to explore how they perceive risk through successive spheres: organizational, physical, digital, and social.

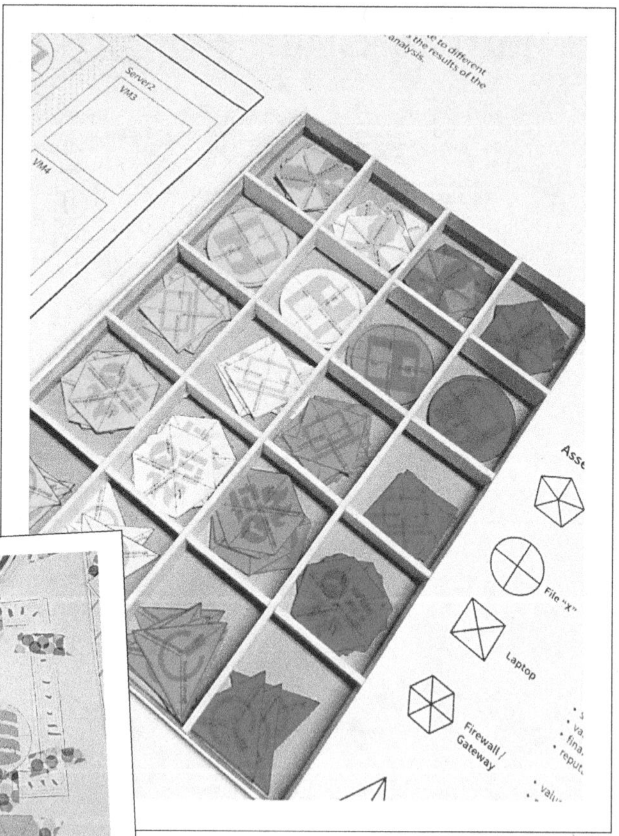

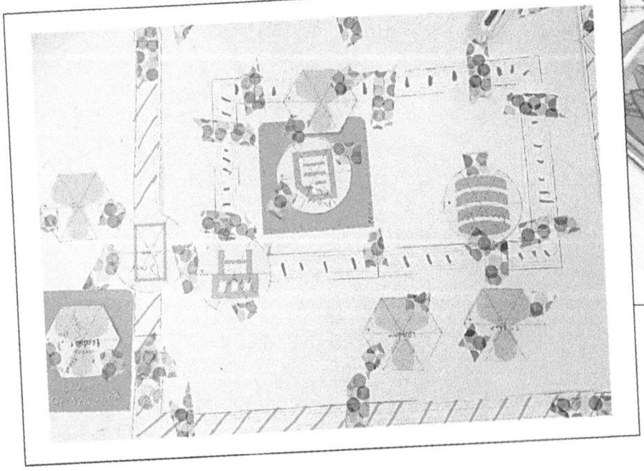

Analogies

Lego simulation

Northwave.(https://northwave-security.com)

Printed with permission from Northwave.

Northwave visualizes business processes using a Lego Train Track. These processes are controlled by employees, protected by Northwave's security operations center and, eventually, disturbed by a malicious party. The derailment of the company or business process through a hack is a visual representation of a cyber security incident.

17. Visualizing with metaphors

A metaphor is a way of describing something by relating it to something else. It usually refers to common characteristics between the two. This common characteristic can relate to a physical similarity, its history, or its constituent parts. There are endless possibilities to bring cybersecurity concepts to life through metaphors.

 **Iceberg**	 **Pyramid**	 **Temple**
 **Mountain**	 **Fort/tower**	 **Sandwich**
 **Gears**	 **Clock**	 **Conveyor belt**
 **Funnel**	 **Bridge**	 **Ecosystem**
 **Puzzle**	 **Balance**	 **Gauge diagram**

References to illustrations other than the ones presented in this chapter are:

Iceberg: cybersecurity threats[179], internal audit and cybersecurity[180], financial losses from cyberattacks[181], cyber incident costs[182]
Ecosystem: cybercrime[183]
Funnel: security logs[184]
Sandwich: identity and access management[185]
Balance: convenience and security[186]
Fort/tower: the top cyber threats[187]

Temple

Is your C-Suite Prepared for the Cyber Future?
J. Lowe and S. Duffy, Electric Energy Online[188]
Printed with permission from Electric Energy Online

This picture is used to explain that C-suite leaders in energy should look beyond just IT and operations to enhance their organization's security posture. The temple is a metaphor where five pillars and the three-layered foundation hold up the roof that is 'trust'. If one of the pillars is unstable, the roof might collapse.

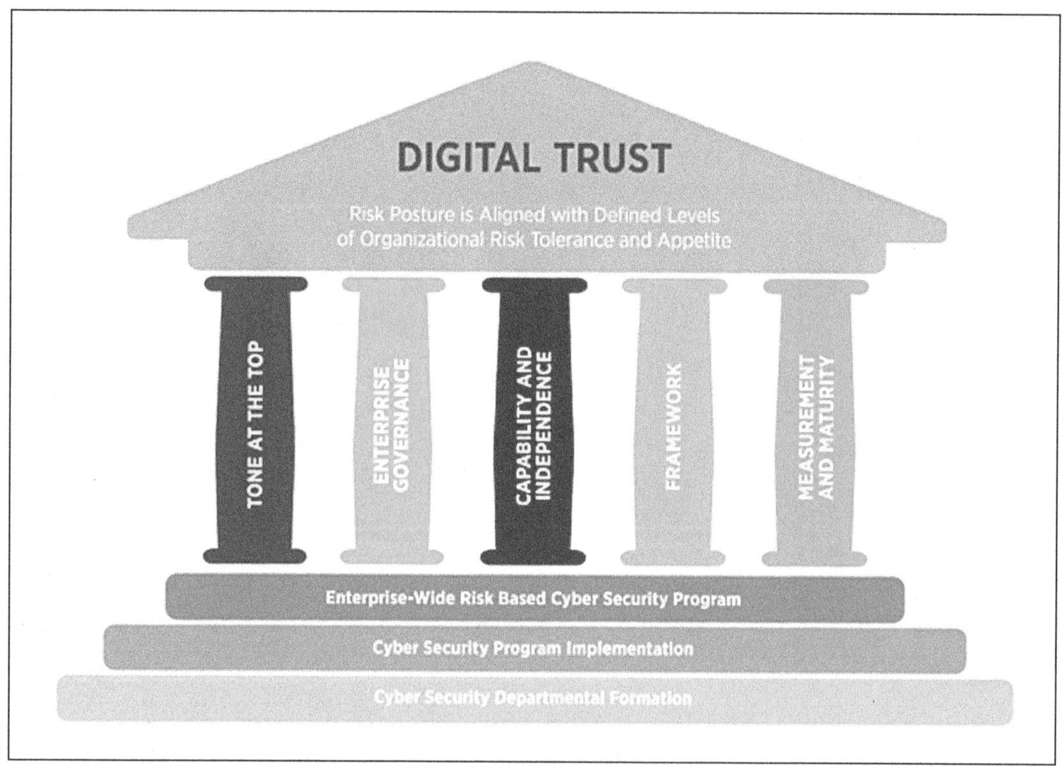

Clock

Around the clock cybersecurity

John McRea, Neuways[189]

Picture used with permission

The clock represents the 12 steps to improve your organisation's security, recommended in a commercial blog.

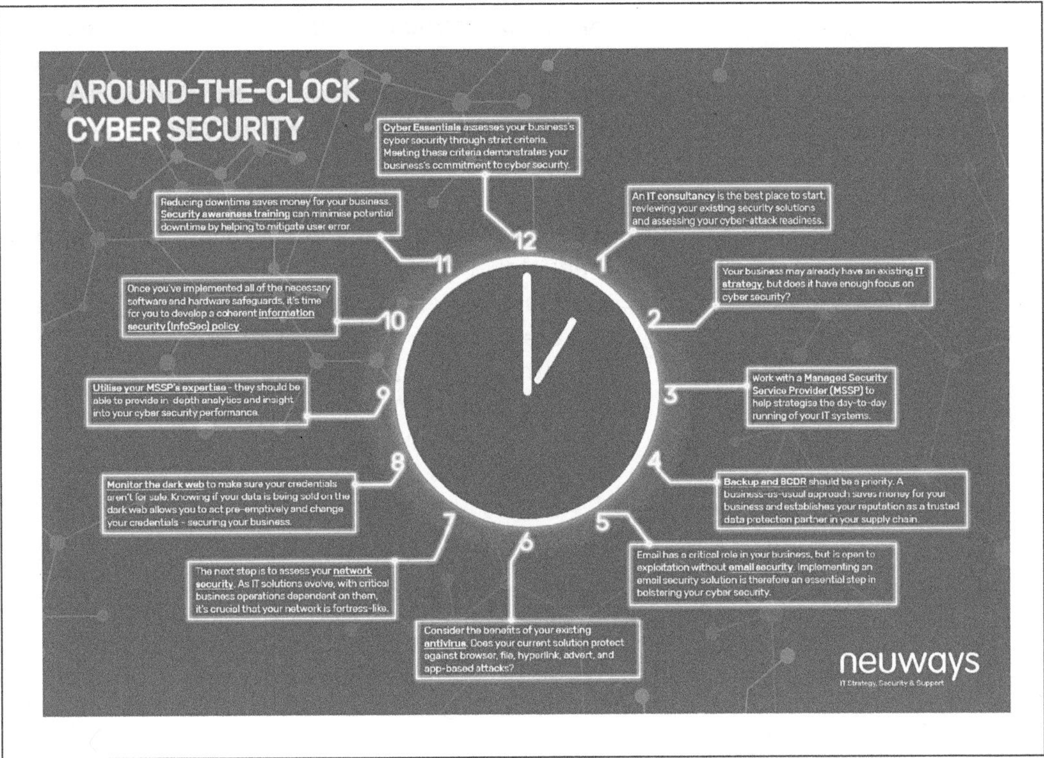

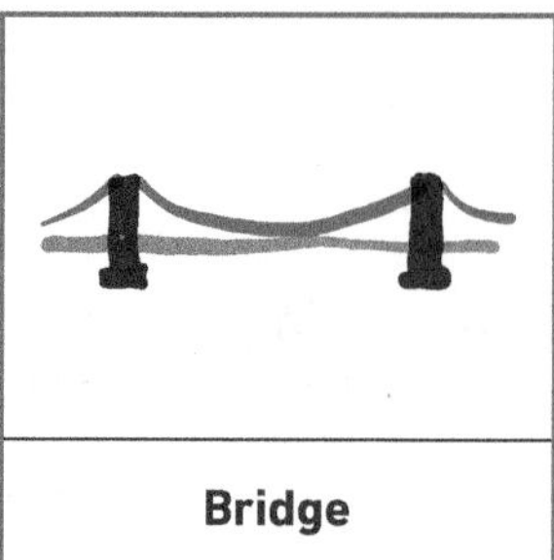

Cybersecurity skills gap?

Nicole van Deursen, created with Piktochart

Based on data from [165]

This infographic is based on a content analysis of 52 job vacancies in The Netherlands in spring 2018. With all the talk about the shortage of cybersecurity professionals, the results of this analysis suggest that it's the businesses own fault. Many of the job descriptions were written in confusing language ('your job is to level the management'); requiring certifications that do not exist (such as: 'you are ISO 27001 certified' -this certificate is for organisations, not for people-); or they demand impossible combinations (a CISSP certificate -that requires >5 years of experience- combined with less than 2 years of working experience). Almost half of the employers do not really care about the level of education the candidate has completed. They write: 'you have a vocational or university degree'. That's not really encouraging for university graduates... Worst of all, there is no demand for the more experienced professional. 67% of employers are not interested in professionals with more than 5 years of experience.

Cybersecurity skills gap

Vacancy texts contribute to the problem

Demotivating requirements

67%

ask for <5 years of experience

(whilst the largest cohort of the working population in NL >35 years of age)

Insincere language

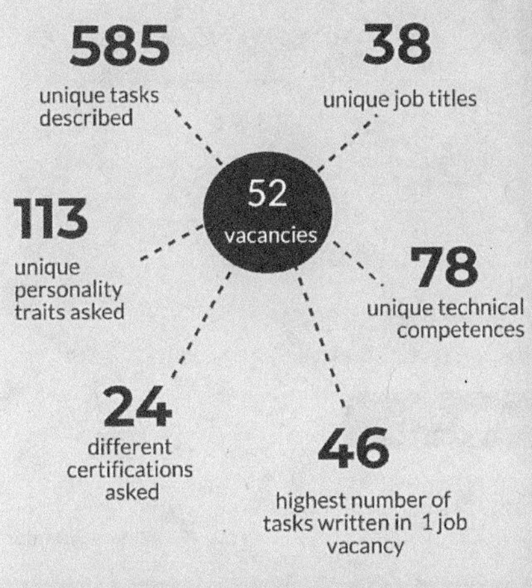

Medium or higher vocational education

Higher vocational education or university

University or CISSP

CISA

CEH

CISSP

CIPP/e

CISM

CRISC

73%

use **confusing language**

23%

require non-existing skills and certificates

48% of employers state that different levels of education and certificates are interchangeable

Content-analysis of 52 vacancies for information security jobs. Nicole van Deursen. May 2018

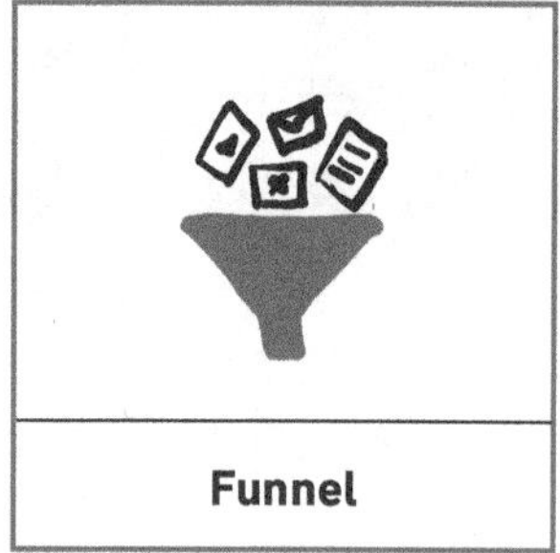

Data Science Hunting Funnel

Austin Taylor[190]

Copyright: permission from author

The Data Science Hunting Funnel was created to illustrate a workflow for security researchers and data scientist to help reduce their dataset and have the best likelihood of identifying malicious traffic and also attempt to set expectations.

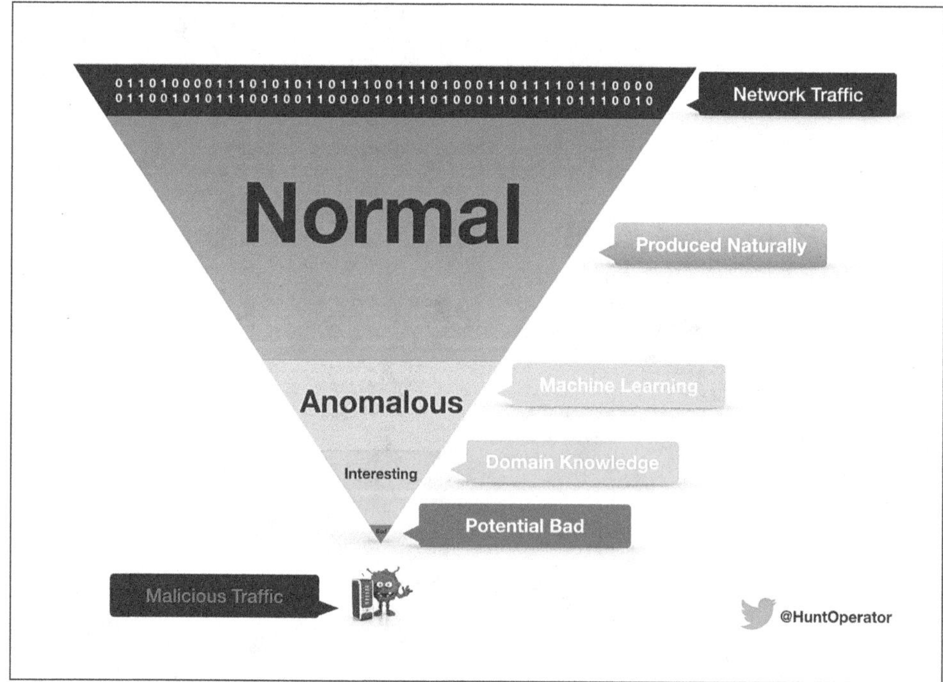

18. Visualizing with style and colour

Sometimes you may not have the option to create pictures and you are in a situation that you just have to go forward with a lengthy textual document. There is a lot you can do to make a text attractive. By using colour, different fonts and font sizes, you can draw attention to the important parts. This will help your reader to quickly scan the document and see the keywords in an instant. The designers that worked on this book also used typography and colour options to make this book more visual.

Also, tables benefit from colour and style such as coloured dots and labels or bold headings.

"OPINION"	typo gra ph y	
 **Quotes**	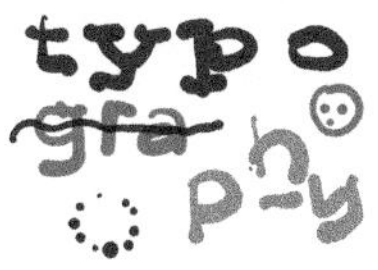 **Typography**	 **Word cloud**
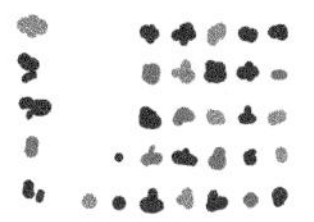		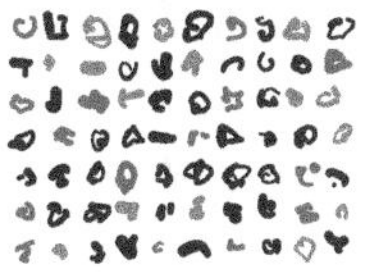
Indicator dots	**Isotypes**	**Icons**
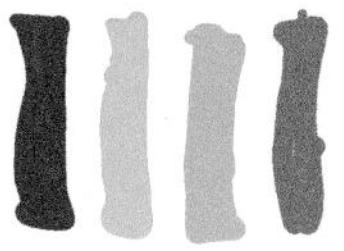		
Colour	**Shading**	**Patterns**

References to illustrations other than the ones presented in this chapter are:

Indicator dots: cybersecurity rating[156]
Icons: risk assessment[127], privacy icons [191–193]
Typography: overview of online threats[194]
Colour: Commonly used TCP / UDP port numbers[195]

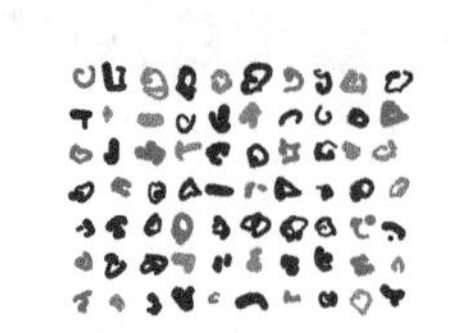

Icons

NCSC Glossary

National Cyber Security Centre, UK[148]

National Cyber Security Centre

NCSC Glossary

This glossary explains some common words and phrases relating to cyber security, originally published via the **@NCSC** Twitter channel throughout December. The NCSC is working to demystify the jargon used within the cyber industry. For an up-to-date list, please visit **www.ncsc.gov.uk/glossary**.

Antivirus
Software that is designed to detect, stop and remove viruses and other kinds of malicious software.

Cyber security
The protection of devices, services and networks - and the information on them - from theft or damage.

Firewall

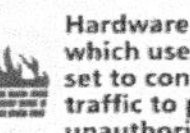

Hardware or software which uses a defined rule set to constrain network traffic to prevent unauthorised access to (or from) a network.

Ransomware
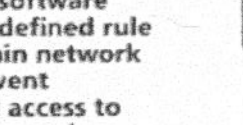

Malicious software that makes data or systems unusable until the victim makes a payment.

Two-factor authentication (2FA)
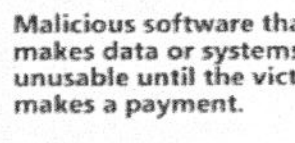
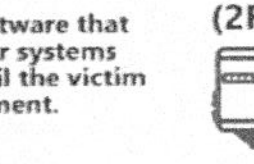

The use of two different components to verify a user's claimed identity. Also known as multi-factor authentication.

Botnet
A network of infected devices, connected to the Internet, used to commit co-ordinated cyber attacks without their owners' knowledge.

Denial of Service (DoS)
When legitimate users are denied access to computer services (or resources), usually by overloading the service with requests.

Internet of Things (IoT)
Refers to the ability of everyday objects (rather than computers and devices) to connect to the Internet. Examples include kettles, fridges and televisions.

Software as a Service (SaaS)
Describes a business model where consumers access centrally-hosted software applications over the Internet.

Water-holing (watering hole attack)
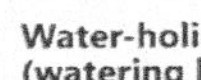
Setting up a fake website (or compromising a real one) in order to exploit visiting users.

Bring your own device (BYOD)
An organisation's strategy or policy that allows employees to use their own personal devices for work purposes.

Digital footprint
A 'footprint' of digital information that a user's online activity leaves behind.

Macro
A small program that can automate tasks in applications (such as Microsoft Office) which attackers can use to gain access to (or harm) a system.

Social engineering

Manipulating people into carrying out specific actions, or divulging information, that's of use to an attacker.

Whaling
Highly targeted phishing attacks (masquerading as legitimate emails) that are aimed at senior executives.

Cloud
Where shared compute and storage resources are accessed as a service (usually online), instead of hosted locally on physical services.

Encryption
A mathematical function that protects information by making it unreadable by everyone except those with the key to decode it.

Patching
Applying updates to firmware or software to improve security and/or enhance functionality.

Spear-phishing
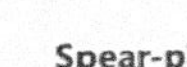

A more targeted form of phishing, where the email is designed to look like it's from a person the recipient knows and/or trusts.

Whitelisting
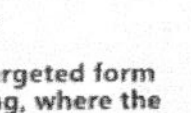
Authorising approved applications for use within organisations in order to protect systems from potentially harmful applications.

Cyber attack
Malicious attempts to damage, disrupt or gain unauthorised access to computer systems, networks or devices, via cyber means.

End user device
Collective term to describe modern smartphones, laptops and tablets that connect to an organisation's network.

Phishing
Untargeted, mass emails sent to many people asking for sensitive information (such as bank details) or encouraging them to visit a fake website.

Trojan
A type of malware or virus disguised as legitimate software, that is used to hack into the victim's computer.

Zero-day

Recently discovered vulnerabilities (or bugs), not yet known to vendors or antivirus companies, that hackers can exploit.

For more information go to www.ncsc.gov.uk @ncsc

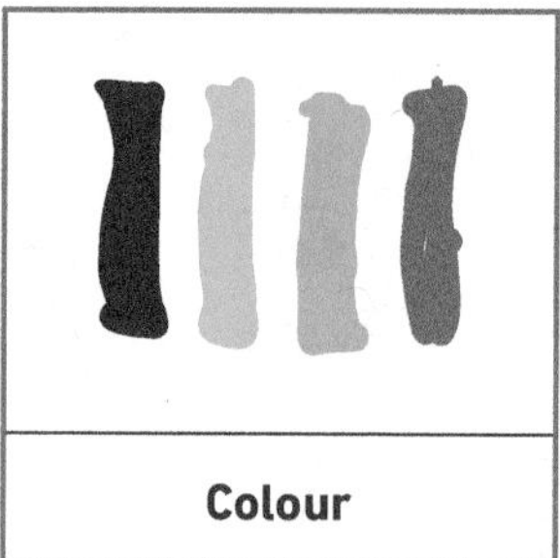

Colour

Risk Matrix
Nicole van Deursen

Many organisations use a variant of this risk matrix with colour codes to visualize the severity of potential risks. The matrix can be used in different situations, amongst which are risk assessment sessions or management reports.

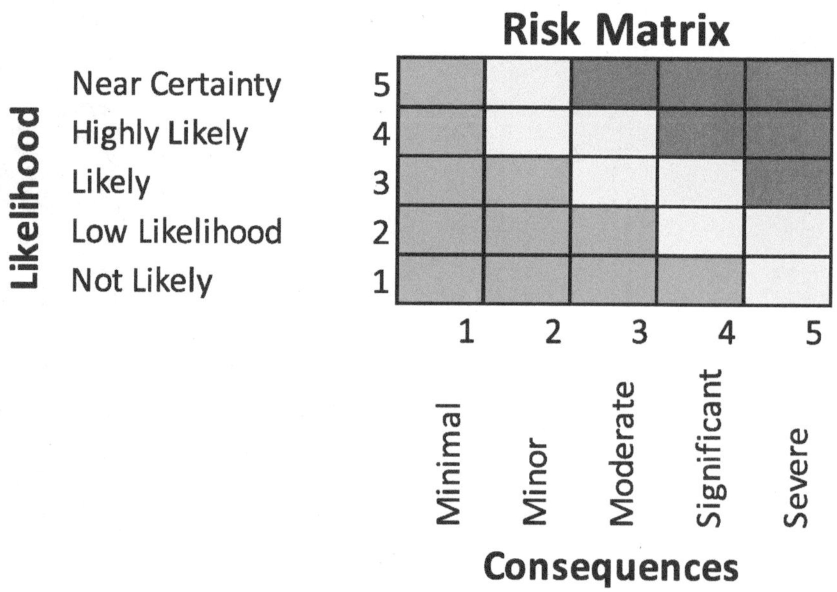

Risk Matrix

Likelihood

Near Certainty 5
Highly Likely 4
Likely 3
Low Likelihood 2
Not Likely 1

1 2 3 4 5

Minimal Minor Moderate Significant Severe

Consequences

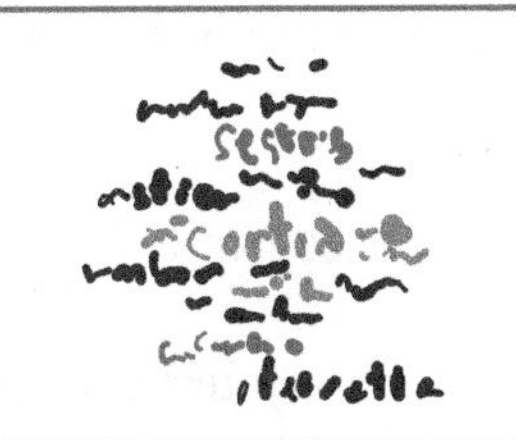

Word cloud

Word cloud of the ISO 27001

Nicole van Deursen

A word cloud is a collection words represented in different sizes. The bigger and bolder the word appears, the more often it's mentioned within a given text. This word cloud is based on the text of the ISO 27001 standard for information security.

responsibilities assets
business Objective(s)
organization
policy
requirements assessment
review
results users system
process procedures action
Control systems
effectiveness unauthorized
facilities management
activities risks Standard
appropriate access
International
ensure security
information

Isotypes

Social engineering and how to win the battle for trust (part of larger infographic).
OneSpan [196]

This illustration is part of a larger infographic on social engineering. The infographic displays data as isotypes: pictograms that represent quantities, not by size of the pictogram but by a greater number of the same-sized pictogram.

Isotypes come from the theories of Otto Neurath (1882–1945), a Viennese philosopher, economist and social scientist. He invented this system to pictorially organise statistics.

So•cial En•gin•eer•ing

noun. ...the art of manipulating people into performing actions or ***divulging confidential information.***

TYPES OF ATTACKS

PHISHING

Email sent under false pretenses to trick users into supplying attackers with their login credentials

Types of Accounts Phishers Target

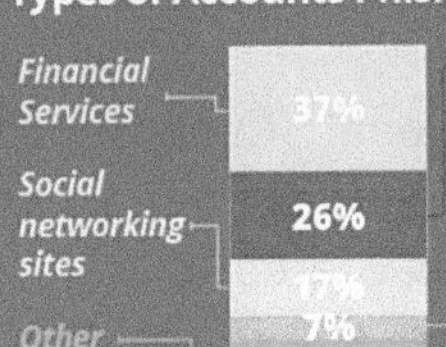

SPEARPHISHING

A targeted phishing email

91% of the most advanced attacks start with a spearphishing email

VISHING (Voice Phishing)

Calling a target pretending to be a person of authority, such as an IT supervisor, to pump someone for credentials or important information

UK banks alone lost £21 million from vishing attacks in 2014

SMISHING (SMS Phishing)

Phishing messages sent through text messages rather than email

200 million SMiShing messages are sent worldwide every day

= 10 Million SMiSh messages

MINING SOCIAL MEDIA

Learning more about targeted people through social media in order to build better phishing lures

Between **52 million and 97 million Facebook accounts** are fake or duplicate accounts

MAN-IN-THE-MIDDLE ATTACKS

The attacker impersonates a company by hijacking an SSL connection between a browser and legitimate web server by exploiting server-side vulnerability.

Just a single rogue DNS attack in 2014 targeted all of the customers at over **70 financial institutions.**

MAN-IN-THE-BROWSER ATTACKS

Same principle as Man-in-the-Middle, only exploiting vulnerabilities in the browser itself

90% of enterprises are exposed to man-in-the-browser attacks

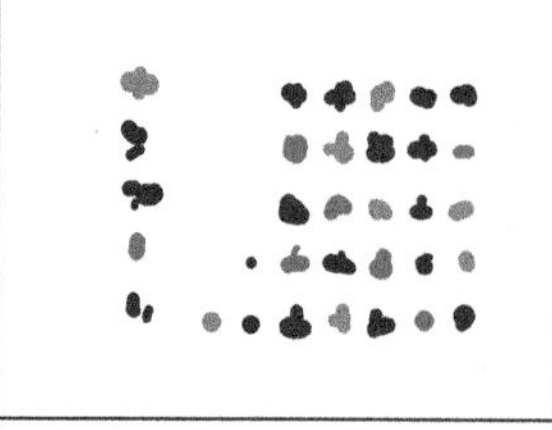

Indicator dots

The Global Cybersecurity Index 2017

International Telecommunications Union[196b].

Copyright: printed with permission from the International Telecommunications Union.

The Global Cybersecurity Index (GCI) follows from a survey that measures the commitment of Member States to cybersecurity in order to raise awareness. 134 Member States responded to the survey throughout 2016. The 2017 publication of the GCI shows level of commitment to cybersecurity of countries around the world. The level of commitment is displayed from green (highest) to red (lowest). This illustration is an easy-to-make and easy-to-understand example of the use of indicator dots to visualize information.

	Cybercriminal legislation	Cybersecurity legislation	Cybersecurity training	**LEGAL MEASURES**	National CERT/CIRT/CSIRT	Government CERT/CIRT/CSIRT	Sectoral CERT/CIRT/CSIRT	Standards for organizations	Standards for professionals	Child online protection	**TECHNICAL MEASURES**	Strategy	Responsible agency	Cybersecurity metrics	**ORGANIZATIONAL MEASURES**	Standardization bodies	Cybersecurity good practices	R&D programmes	Public awareness campaigns	Professional training courses	Education programmes	Incentive mechanisms	Home-grown industry	**CAPACITY BUILDING**	Bilateral agreements	Multilateral agreements	International participation	Public private partnerships	Interagency partnerships	**COOPERATION**	**GCI**
Albania																															
Andorra																															
Austria																															
Belgium																															
Bosnia and Herzegovina																															
Bulgaria																															
Croatia																															
Cyprus																															
Czech Republic																															
Denmark																															
Estonia																															
Finland																															
France																															
Germany																															
Greece																															
Hungary																															
Iceland																															
Ireland																															
Israel																															
Italy																															
Latvia																															
Liechtenstein																															
Lithuania																															
Luxembourg																															
Malta																															
Monaco																															
Montenegro																															
Netherlands																															
Norway																															
Poland																															
Portugal																															
Romania																															
San Marino																															
Serbia																															
Slovakia																															
Slovenia																															
Spain																															
Sweden																															
Switzerland																															
The former Yugoslav Republic of Macedonia																															
Turkey																															
United Kingdom																															

19. Infographics

Infographics show both quantitative data (statistics) and qualitative data (knowledge). This causes some discussion around the exact definition of an infographic. Some state that an infographic is a larger graphic design that combines data visualizations, illustrations, text, and images together into a format that tells a complete story [197,198]. An infographic in this perspective is a story with a header, a middle, and a conclusion or call to action. A single data visualization by itself is not an infographic. Neither is a number nor percentage with a picture next to it. Others state that there are many different forms of infographics possible, ranging from an infographic resume to informational infographics with mostly text. In such a production, the 'data' can be text and as long as its function is clear it counts as an infographic.

An infographic can be an effective means to inform, to convince, to tell a story, or to show an analysis. As long as your audience connects with the infographic, it does not really matter if the design is 'correct'. For that purpose, it is important to test your design on your audience to see if they respond to it as you would expect. Do they understand the message, do they feel it, learn something, or are they compelled into doing something? Learn from their feedback and improve the design. The purpose is to make meaningful infographics. Well-designed infographics are easier for many readers to process cognitively than text-only documents. When published online, they are easy to share within social networks and have the potential to go viral.

The aim of an infographic can be:

To instruct: an infographic helps to explain how to do something or how to use a product.

To inform: it can be used to present research results or to raise awareness for a certain topic.

To persuade: use an infographic as a call to action, to invite people to join, buy, or change something.

To tell a story: an infographic can tell the story through a timeline of events or actions; what happened to someone, how they did it.

To analyse: to demonstrate the root cause of an event, to deduct a sequence of events, and to show it happened.

To compare two things: when comparing two products, best practices, or events over the past two years.

Over the years I have collected many examples of infographics on Pinterest. I invite you to join me and grow this collection together: https://www.pinterest.co.uk/nicole9319/information-security-infographics/.

Introducing the psychology of passwords
LastPass[199]
Copyright: permission from LastPass

This infographic shows the results of a survey by LastPass and Lab42. The survey about attitudes and behaviours around password security found that although people know what safe passwords should be, they tend to ignore this knowledge in favour of using easy-to-remember passwords, because the fear of forgetting is stronger than the fear of being hacked. Furthermore, the personality traits that normally define us seem to have little bearing on our poor behaviour but do help people rationalize it.

THIS IS YOUR BRAIN ON PASSWORDS

Does the theory of cognitive dissonance also apply to our digital behavior? **You know it's bad for you, but you do it anyway.** Although major breaches continue to make headlines, we have not changed how we protect our digital life.

Cognitive Dissonance is the psychological conflict resulting from an individual performing an action that is contradictory to their beliefs, idea or values.

WHAT YOU KNOW

YOU UNDERSTAND WHAT GOOD PASSWORD BEHAVIOR SHOULD LOOK LIKE...

- **59%** know a secure password is important
- **91%** understand the risk of reusing passwords
- **2/3** are fearful of password hacking
- **75%** consider themselves informed on password best practices
- **72%** consider their passwords sufficient protection for their online information

WHAT YOU DO

...YET YOU CONTINUE TO EXHIBIT POOR PASSWORD HABITS

- **41%** choose a password that is easy to remember
- **61%** use the same or similar passwords
- **55%** do it even though they understand the risk

HOW DOES YOUR GENERATION STACK UP?

Millennials [illegible] Generation X [illegible] Baby Boomers [illegible]

?!

#1 reason people change their passwords is because they forgot it.

THE STRONGEST PASSWORDS ARE CREATED FOR:

69%

43%

31%

20%

WHAT YOU REMEMBER

47% are afraid of forgetting passwords

↳ WOMEN ARE MORE AFRAID THAN MEN

55% ♀ 39% ♂

YOU RELY ON PERSONAL INFORMATION TO CREATE AND REMEMBER YOUR PASSWORDS

EASY-TO-CRACK INFO USED TO CREATE PASSWORDS

- 47% your/family names/initials
- 42% significant dates/numbers
- 26% pets
- 21% birthdays
- 14% hometown
- 13% school name/mascot

WHY YOUR PERSONALITY WILL GET YOU HACKED

When it comes to online security, your personality type does not inform your behavior, but it does reveal how you rationalize your bad password habits.

TYPE A

Bad password behavior in Type A personalities stems from their need to be in control. Even though they reuse passwords, they don't believe they are personally at risk because of their own organized system and proactive efforts.

CONTROL

35% reuse because they want to remember all passwords

DETAIL-ORIENTED

49% have a personal "system" for remembering passwords

DELIBERATE

2/3 are proactive to help keep personal info secure

DRIVEN

86% having a strong password makes you feel like you're protecting yourself and your family

TYPE B

Type B personalities rationalize their bad behavior by convincing themselves that their accounts are of little value to hackers. This enables them to maintain their casual, laid-back attitude toward password security.

NONCHALANT

45% believe your accounts aren't valuable enough to make them worth a hacker's time

LAID BACK

43% prioritize a password that is easy to remember over one that is secure

FLEXIBLE

1/2 feel that you need to limit your online accounts and activities due to fear of a password breach

PREOCCUPIED

86% feel other things outside of a weak password could compromise your online security

DON'T JUST RESET YOUR PASSWORD, RESET YOUR THINKING

Managing your passwords properly can be a quick behavioral adjustment that can yield long-term benefits.

LastPass •••|

REGARDLESS OF YOUR PERSONALITY TYPE, LASTPASS CAN HELP YOU MANAGE YOUR PASSWORDS IN A CONVENIENT AND SECURE WAY.
LASTPASS.COM

20. Cybersecurity art

Cybersecurity is increasingly inspiring artists as a topic for motion pictures, museum exhibitions, and artistic data visualizations. Fortunately, more and more businesses work with designers and data artists to create alternative approaches to awareness campaigns by using different art forms and techniques. Working with trained artists delivers innovative and emotionally engaging approaches to draw attention to the complexity of the field and the impact of technology on society. Art has the strength to express emotions such as fear, confusion, optimism, or enthusiasm. Through art, an audience can understand and feel the limitations and the possibilities of cybersecurity. This chapter highlights some of the initiatives where cybersecurity experts collaborate with professionally trained artists.

20.1 Photography

The first chapter of this book pointed out that the media often pictures cybersecurity experts as shady figures wearing hoodies. In The Netherlands, the photo exhibition called Hackers Handshake brings an honest and more diverse image to the public. Hackers Handshake is a project by Tobias Groenland[200]. He started to investigate the hackers' scene expecting that they would confirm the stereotypes. In contrast, he discovered a diversity of individuals who have a passion for technology in common. The 16 portraits in the exhibition tell the story of helpful hackers who seek, find, and report vulnerabilities in computer systems in order to improve security. They report their findings to the owners of the vulnerable systems first and then to the hacker community, governments, and businesses that are related to that system, a practice is known as Responsible Disclosure, or Coordinated Vulnerability Disclosure[201].

Between 2014 and 2018 Tobias Groenland and author Chris van 't Hof collected stories of hackers using their skills for good purposes. Chris van 't Hof wrote the texts that tell the stories behind the portraits and published a book called Helpful Hackers[202] that portraits Dutch hackers and the concept of responsible disclosure. The message behind this work not only contributes to a realistic and honest public image of cybersecurity experts, it might also invite young people with advanced technical skills to choose to use their skills for good instead of straying into cybercrime.

Hackers Handshake photo exhibition.
Photographer: Tobias Groenland (https://www.tobiasgroenland.nl).
Image Copyright © Tobias Groenland.

During the photography sessions the photographer requested to think of coding during the shoot. In their minds they might be scanning for vulnerabilities or doing a DDoS attack and as a result they might be revealing something of themselves to the wider audience through their glaze.

20.2 Data art

Data visualization for cybersecurity has further evolved since the publication of Raffael Marty's book on Applied Security Visualization in 2008[121]. We can even see an emerging trend where cybersecurity data turns into art. However, data art is not the same as data visualization. Data art does not seek to present unmistakable evidence and to explain structures from data, but rather creates a visual experience or seeks new methods of presentation[203]. Data artists take the data and create new compositions from it.

Books such as Knowledge is Beautiful and Information is Beautiful and their supporting website[149,204,205], or the collections of Manuel Lima, amongst which are The Book of Trees[206], the Book of Circles[207], and Visual Complexity[208] are inspirational to find new ways to visualize data in beautiful ways, and with many of the examples in these books it's not easy to separate the explanatory data visualization from data art.

Technology businesses collaborate increasingly with artists. Trend Micro commissioned data artists from across the globe to transform real Trend Micro data into works of art[209]. They can be downloaded for free as a screensaver.

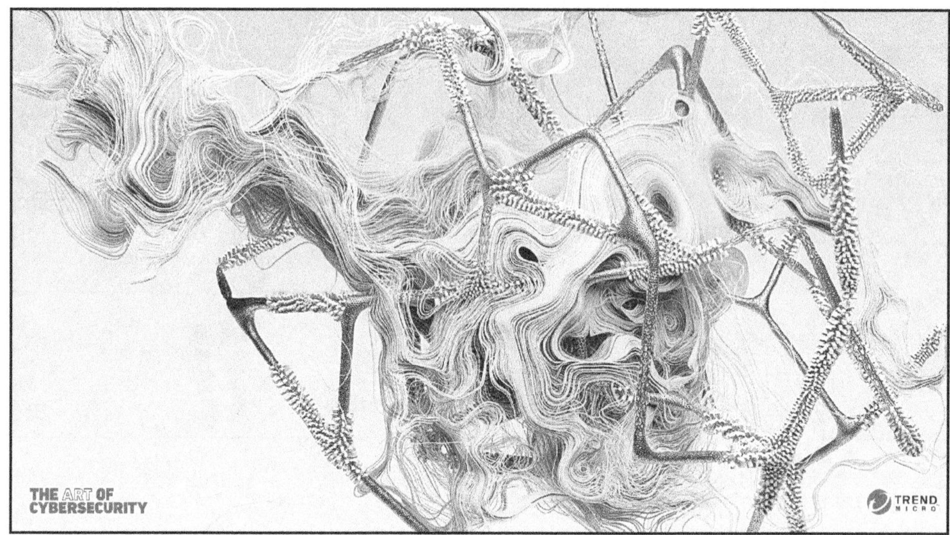

Unknown Threats Detected & Blocked Over Time by Brendan Dawes [210].

Endpoint Threats Detected & Blocked Over Time by Stefanie Posavec [211].

20.3 Malware exhibitions

Computer viruses have become part of our daily digital life and the design and creativity behind malware is related to digital culture. In this light, the activity of programming a virus is seen as an art form and cultural practice that, in the words of Jussi Parikka, 'refashions our sense of security and self'[212]. Malware is much more than just a security issue. It is a way to turn a real situation into a lie and an invention of an accidental situation that turns out into the advantage of the creator. The artistic value is placed in the paradox of the beautiful complexity behind the malware and the damage it can cause.

An overview of different malware programs can be found at The Malware Museum. The Malware Museum is an online collection of malware examples at the Internet Archive[213]. It shows a selection of malware from the 1980s and 1990s on home computers.

The most famous and more recent examples of malware were displayed in 2019 in The Netherlands in an exhibition called Malware: symptoms of viral infection, curated by Bas van de Poel and Marina Otero Verzier[214]. The exhibition showed images of the malware and engaged the audience in questions about safety, war fare and geopolitics. As many viruses cannot be seen, the curators created visualizations to match the stories behind the creators of the virus or those of the victims.

An earlier exhibition to feature viruses and malware was I Love You [rev. eng], conceived and presented by the digitalcraft.org Kulturbüro, curated by Franziska Nori, and first shown in Frankfurt in 2002. The exhibition was later expanded and travelled through different countries until 2006[215]. The exhibition was divided into four investigative areas - political, cultural, technical and historical. It included for visitors to:

- Force computers to crash
- Experience a global virus outbreak in real time via a 3D world
- View security concepts and methods for preventing global network attacks
- Witness computer viruses as works of art
- See films by hackers on their subculture
- Learn about programming languages as the material for contemporary poetry
- Compare experimental literature and code poetry

20.4 Immersive art

Immersive art is a form of art where the viewer becomes part of the artwork. The viewer can walk within the installation or can interact with it. They are often multi-sensory experiences that include sound, scent, touch, and vision.

Immersive art can make the spectator feel like they are part of something big and even infinite. They appear to be in the middle of the elements of time and space. This art form suits the perception of the infinity of internet and the connections of a single person with countless numbers of networks.

Immersive art exhibitions with a technology theme often are inspired by big data, neural networks, artificial intelligence[216], and machine learning. Some of them even use machine learning algorithms on a database with images to display those images in multiple dimensions in a way that the algorithm decides[217].

Poetic AI by OUCHHH STUDIO.

Photo: author's private collection. Taken at Atelier des Lumières, Paris (2018).

Part IV

Workbook

21. Introduction

Cybersecurity is an exciting field to work in and offers many career paths for people from different disciplines. Professionals working in cybersecurity have many complex problems to solve and, therefore, they need to collaborate amongst each other and with professionals from other fields. This means that they have to be able to translate technical concepts and jargon into a language that their colleagues and audience understand.

Visual communication is a tool that will help to advance the cybersecurity field in many ways:

- It improves collaboration.
- It helps to explain complex topics.
- It makes cybersecurity understandable.
- It radiates enthusiasm.
- It invites to share stories.
- It engages end-users.

With improved communication, cybersecurity professionals will be able to design better security products and services in multi-disciplinary teams. Visual tools are fun to share with others. When we share our visual stories, we all become advocates of cybersecurity. If we are enthusiastic, others will follow.

Unfortunately, enthusiasm alone will not give you the budget to go on a visual design course, hire visual facilitators, or to employ professional designers. You will have to make the right stakeholders enthusiastic as well, so that they will spread this enthusiasm and help others to get the right mindset for visual communication.

The first activity in this workbook, drawing a stakeholder map, is a practise in visualization as well as an assignment to work with stakeholders that can help you to achieve support for visual communication in your organization.

The second activity, a visual dictionary, will help you to start using simple visuals in your communication as well as achieving a shared language in your organization.

The steps in each activity are based on the concepts of design thinking. The steps are not necessarily chronological: they can occur in parallel and can be repeated iteratively. Our work environment is dynamic, the circumstances change continuously, and we need to always be ready to learn, listen, and improve our visual tools.

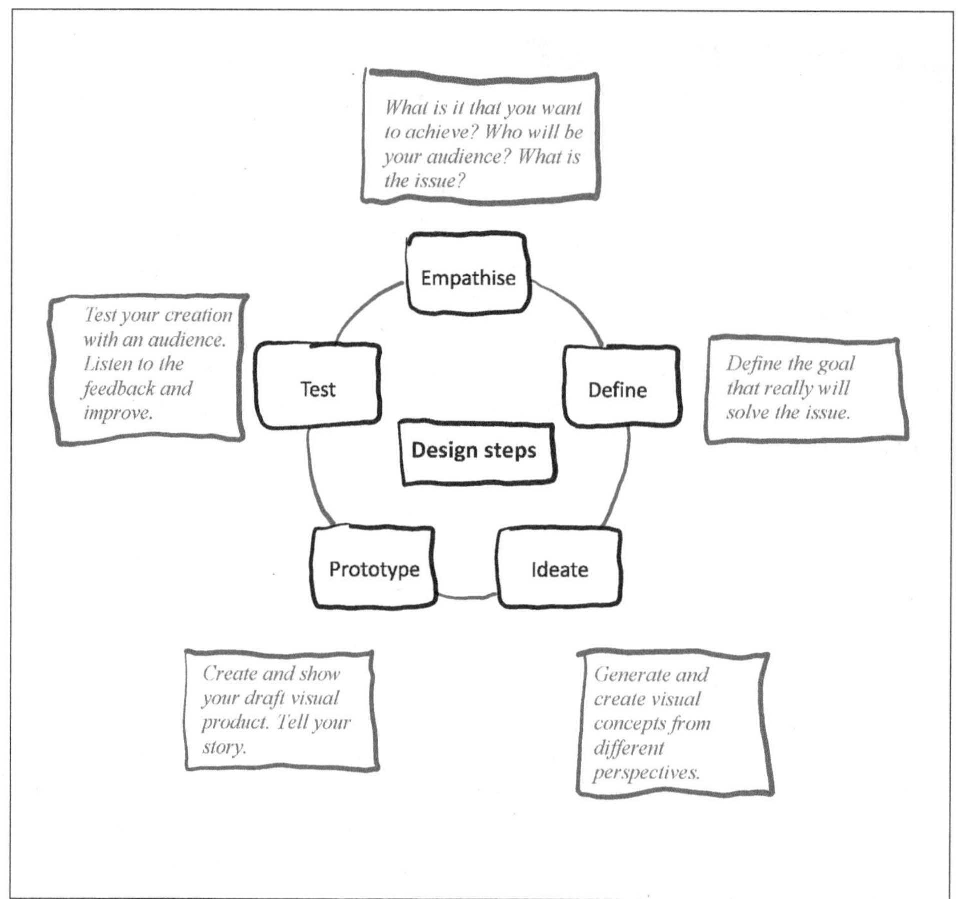
What is it that you want to achieve? Who will be your audience? What is the issue?
Empathise
Test your creation with an audience. Listen to the feedback and improve.
Test
Define
Define the goal that really will solve the issue.
Design steps
Prototype
Ideate
Create and show your draft visual product. Tell your story.
Generate and create visual concepts from different perspectives.

22. Activity 1: Stakeholder map

Assignment

You are a cybersecurity professional who sees the advantages of visual communication. You are not great at drawing pictures or at explaining complex topics. You have no budget of your own and your manager is only interested in getting the work done without any frills. She says that visuals do not fit into the company culture. You think that you are not able to explain the advantage of visual communication good enough to persuade your manager to send you on a visual design training course or to hire a visual facilitator.

Remembering the sections from this book about stakeholders, advocates, and how people are more likely to be persuaded if their peer groups recommend it, you decide to find stakeholders that have the power to influence your goal. Even if these people do not have any formal relation to your work, they may influence the success of your mission. A stakeholder map will help you to get insight into all people that are interested in visual communication and all people that formally and informally influence your goal.

Step 1: Understanding

If you know some co-workers who are also interested in getting more visual, invite them to collaborate with you. Together you can achieve more. The first challenge is to identify all relevant stakeholders and how they may benefit from visual communication.

With your team, you can think of all the work that you are doing and with whom you are related. Creating a stakeholder map is like creating a map of your own environment and the people that you know. Some questions you could ask are:

1. Who gives you budget?
2. Where does your work come from?
3. With whom do you need to cooperate?
4. Who thinks they are involved (but are not)?

5. Who are involved (but does not act on it)?
6. Who are the people that you need to explain complex topic to?
7. Who may have the resources that you need?
8. Who are the external groups that can help you?
9. Who is really good at visual communication?

The list will give an overview of whom to speak to next. You could ask them what they think about visual communication and how it will help them in their work. Do they experience any problems with the communication with the cybersecurity team? Get to know them to find out how they think that visual communication could help them (after all, you do not want this for yourself, but you want it for others so that they get a better understanding of cybersecurity). Speak to people who are already very visual, even if they are in a completely different line of work. Ask them how they convinced their management. What are the advantages that they experience?

Step 2: Define the goal

Decide how you will use your stakeholder map. Will you use it to explain to your management how complex your environment is to communicate with? Will you keep it to yourself as a map towards influencing 'difficult' stakeholders? Will it be a map supported with data to prove how many stakeholders are with you in this topic? Who will be looking at the map? What other goals could you use it for?

In general, there are three essential goals in stakeholder mapping:

1. Identifying the key stakeholders and their interests (positive or negative) in your goal
2. Assessing the influence and importance of each stakeholder
3. Identifying how best to engage them

Steps 3 and 4: Ideas and prototypes

As steps 3 and 4 are closely related, it makes sense to combine them in this case. With your team, you can start organizing your list of stakeholders. There are different options thinkable:

- Formal versus informal stakeholders
- Internal versus external to the organization
- Individuals versus groups
- Visible (in your team) versus invisible (contributors such as support teams)
- Stakeholders of your stakeholders
- Influencers versus followers

The result may look like a mind map, spider's web, or network diagram showing who is related to who. For instance, suppose you know a person who is a gifted storyteller and always uses that technique in presentations. You are not connected to that person through work and have never met her in person. You can still add this person to your map as an important champion of your goal. If you find out who knows her and how you are connected to these people, they may be able to introduce you. This person may be able to help you influence the people that you need to engage. You can advance the diagram or mind map by adding emoticons, arrows, boundaries, or colours to indicate good/bad relationships, influences, power, and so on.

You could start with a template as below and make changes as you go. Maybe you will invent your own type of map after organizing your stakeholders in different groups and relationship types.

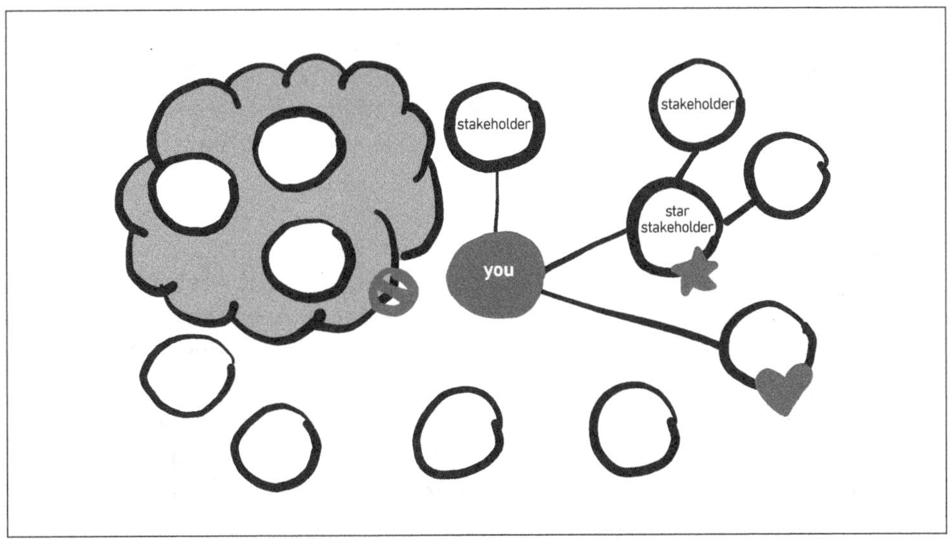

The next step is to plan the communication strategy for each stakeholder. It may help to first organize your stakeholders in a power grid. The grid has four quadrants:

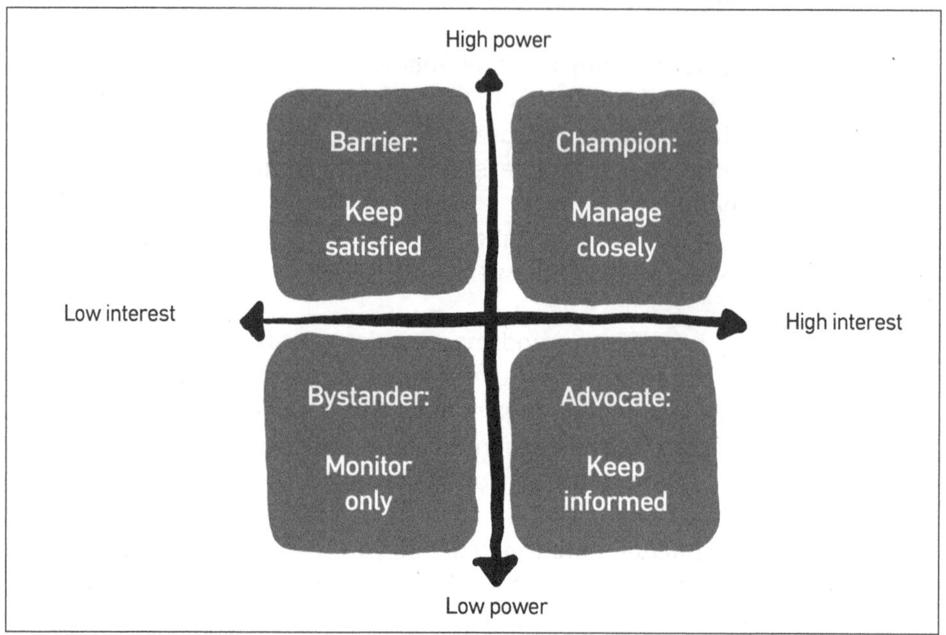

Key players, champions (top right)

This is where to really focus your efforts. These are the people who have the power to support your idea and are interested in visual communication. Invest time to get to know these stakeholders. Find out what matters to them. Keep proactive and collaborative relationships with them. Explain to them what you want to achieve and how you all will benefit from it. These people might have the influence to demonstrate your management the need to invest in visual communication. Give them this book, they may pass it on to their peers in the next group: the barriers.

Barriers: Keep satisfied (top left)

These are the second priority group. They have influence on your goal but less interest in the results. Keep your communication to them brief and to the point. Consult them: their opinion matters and they have the power to help you to be successful. It is also worth to invest time in building relationships with the people who influence these stakeholders.

Advocate: Keep informed (bottom right)

These are less important than the two groups above, but they are interested in visual communication. This group is a potential group of advocates. Share visual messages such as infographics and videos with them. With their enthusiasm, they are likely to send them on to their peers who will send it on and on. One day, those messages might just reach your most powerful stakeholders.

Bystander: Minimal effort (bottom left)

These are the least important stakeholders for your mission right now. Do not waste too much time on them on this topic.

It is easy to draw a power grid, just two lines will do the trick.

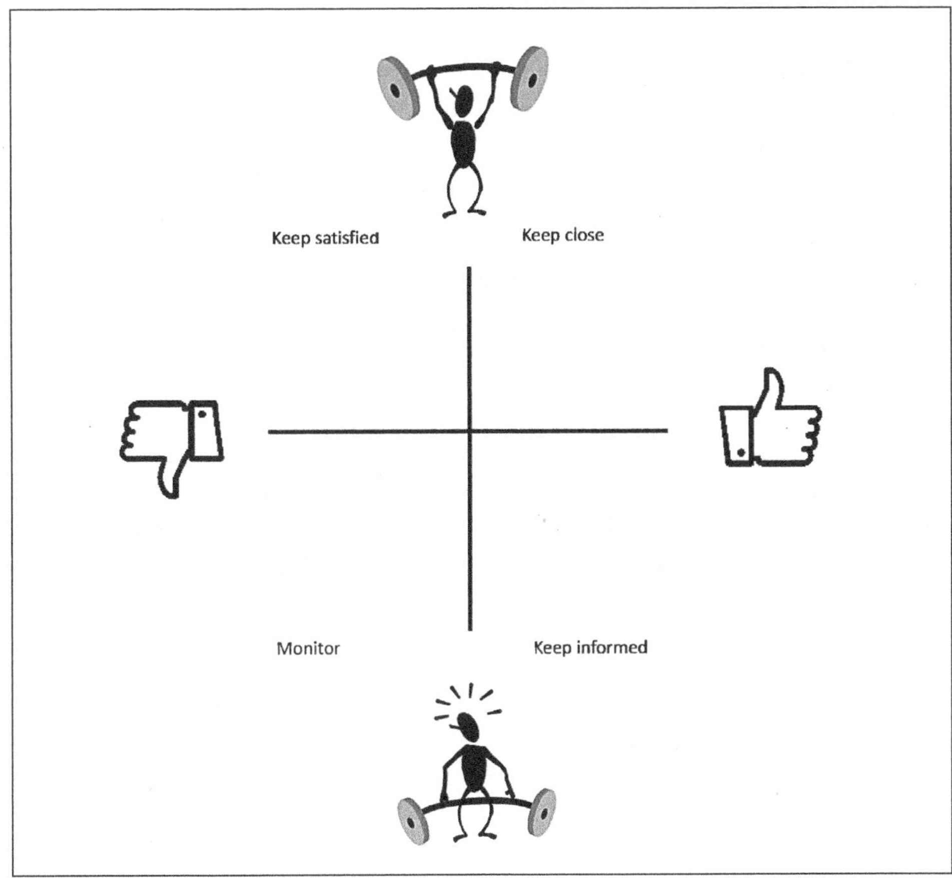

Step 5: Test and improve

Engage with your stakeholders as you planned in the previous steps. Evaluate on a regular basis whether your map is still correct and whether your approach is working. New stakeholders may appear. Some may change roles and lose power or interest.

Do not give up and above all present visually!

23. Activity 2: Visual dictionary

Assignment

Imagine you have just finished your stakeholder map and during the time that you created it, you have met a lot of people who are enthusiastic about visual communication. Your manager is still not convinced. You decide that an example is necessary to support your case. The example you choose is a visual dictionary. Defining and visualizing your cybersecurity abbreviations and jargon is a good topic to start with. Cybersecurity experts use a variety of taxonomies and dictionaries. Different security dictionaries give different definitions to the same jargon. To support a shared language in your organization, you should choose one of those dictionaries and use it consistently. It is a good exercise if you decide to create your own dictionary. It is even better if you create it as a team building exercise, and it is a relatively easy topic to start with. Creating a shared language is important to engage with audiences outside the cybersecurity team and also for collaboration amongst your team of cybersecurity experts.

A dictionary can be supported with icons and symbols. They are so powerful because they

- Work anywhere. A well-designed icon has the potential for communicating across language barriers.
- Cut through the clutter. Between all the words in print and online, an icon stands out to help readers navigate, guiding them to various kinds of content.
- Create recall. Some organizations use icons to represent key information, like company values. Icons or symbols serve as 'handles' that employees can grasp when trying to recall information.
- Invite interaction. Icons on websites draw attention and invite the visitor to look for more information.

Step 1: Understanding and empathizing

Call in your team. Which cybersecurity terms and abbreviations do you use? Does everybody know how to explain what these words mean? Together, make a list of jargon used in cybersecurity documents and presentation slides. What are the words that the experts throw at the people outside the team?

With this list, go and interview some people outside the security department. Have they ever heard these words? Do they know what they mean? Are these words relevant for their job? Where would they go to find out the meaning of a word they do not know? How much understanding of cybersecurity concepts do they have? You can organize your findings in a list, in a 2 × 2 matrix, or write each finding on a post-it and stick it on the wall.

Going well: Jargon easy to explain and well understood	**Could be better:** Jargon difficult to explain and hardly understood
What goes wrong: Jargon easy to explain and wrongly understood	**Miracles:** Jargon difficult to explain and well understood

Step 2: Define the goal

Define the goal of your visual dictionary. Will it be used for reference only or will it be used as a part of cybersecurity training? Is it for internal use only or will you publish it to third parties as well? Will it be presented in an app or online to explain privacy and security settings? Multiple goals might involve multiple types of audiences that, in turn, may require different communication styles and channels.

Step 3: Let your ideas out!

Invite your team again and this time also invite representatives of your potential audience. Make sure you prepare the room. There must be enough space to walk around a little and to stick pictures to a wall. You will need paper, markers, sticky notes, tape, scissors, and so on.

Together, think about possible options for the dictionary. What style of language will work? Formal or informal? Should you just give a definition or does the audience appreciate examples as well?

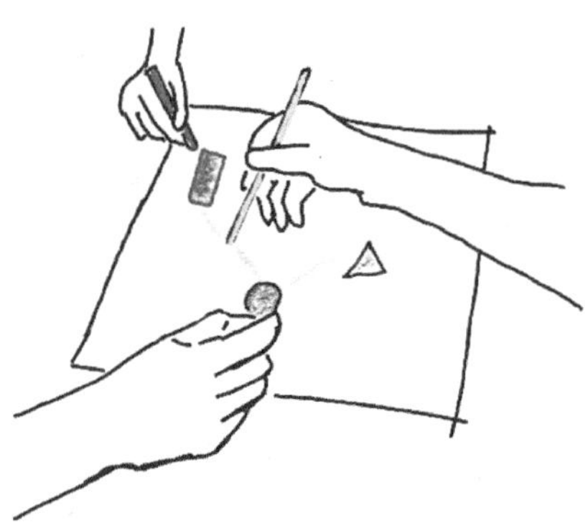

Adopt a beginner's mindset. Start sketching with pen and paper. Make a list of definitions and start drawing icons. It may be difficult in the beginning: visual thinking is like muscle: it needs warming-up. If you can find inspiration in existing definitions and icons in other dictionaries, it is alright to copy them for the moment. At this point, it does not matter; they are likely to change or be dropped during the process (if not: check the copyright and pay the licence if necessary). You can find inspiration in icon libraries such as [218,219].

Here is a list of useful books that will help you to get started with drawing:

Dan Roam. *Show and Tell*. Penguin, 2014 [220]

Sunni Brown. *The Doodle Revolution*. Penguin, 2014[31]

Willemien Brand. *Visual Thinking*. BIS Publishers, 2017[221]

Working on an icon library for secure software development.
Photo: author's private collection.

Step 4 and Step 5: Prototype and test

Cut out your team's doodles and pictures and start organizing them on a paper next to the definitions of the jargon. This creates a prototype of your visual library. You can test the dictionary on different people in your organization. What is their opinion? Do they understand the icons and the definitions? You could maybe re-visit the people that you met during Step 1 and ask them what they think. With this feedback, you can improve your dictionary.

When your team is happy with the results, it is time to go digital. You can re-draw your icons yourself with your tool of preference.

It is important to publish your dictionary for everybody to find. And use it consistently: use your icons in your documents, presentations, and on your webpages. The more you use them, the more feedback you will receive. If they are not clear, get outdated, or become a topic of discussion: go back to Step 1. Change them. Improve where necessary. Remove words that you do not use any more and add new ones that have become important over time.

The more you use the icons, the more people start recognizing cybersecurity messages. You may win over a few new stakeholders. Use the positive reactions in discussions with your manager. Maybe the positive response will finally convince her that visual communication helps to engage other people with cybersecurity.

A final note

I started this book with a call for change. The image of cybersecurity can do with a positive change: we need more people in our profession, we need collaboration with other disciplines, and we should stop treating computer users as the enemy. By pointing out to you the relevant communication theory and models, I hope that you can find some anchors towards understanding how your audience thinks and behaves in their environment. I am grateful to the people I met throughout the process of creating this book and I cannot thank them enough for going to great lengths to help me collecting the images.

Hopefully you'll find the examples in the book inspirational and perhaps you will send me more examples and case studies to include in future work. Please create more visuals, share them, and enjoy. Cybersecurity is cool. Show it to the world!

Bibliography

1. Morgan S. Cybersecurity Jobs Report 2018-2021. Cybercrime Magazine. https://cybersecurityventures.com/jobs/. Published February 23, 2018. Accessed February 19, 2019.
2. Dunn Cavelty M, Van der Vlugt RA. A tale of two cities: or how the wrong metaphors lead to less security. *Georgetown Journal of international affairs, suppl international engagement on cyber V.* 2015;(Fall 2015):21-29.
3. Haney JM, Lutters WG. "It's Scary...It's Confusing...It's Dull": How Cybersecurity Advocates Overcome Negative Perceptions of Security. In: *Fourteenth Symposium on Usable Privacy and Security (SOUPS 2018).* Baltimore, MD: USENIX Association; 2018:411–425. https://www.usenix.org/conference/soups2018/presentation/haney-perceptions.
4. SANS. Security Awareness Compliance Requirements. October 2017. https://www.sans.org/sites/default/files/2017-12/sans-compliance-requirements.pdf.
5. awareness | Definition of awareness in English by Oxford Dictionaries. Oxford Dictionaries | English. https://en.oxforddictionaries.com/definition/awareness. Accessed February 19, 2019.
6. Kaur J, Mustafa N. Examining the effects of knowledge, attitude and behaviour on information security awareness: A case on SME. In: *2013 International Conference on Research and Innovation in Information Systems (ICRIIS).*; 2013:286-290. doi:10.1109/ICRIIS.2013.6716723
7. Chandarman R, Van Niekerk B. Students' cybersecurity awareness at a private tertiary educational institution. *The African Journal of Information and Communication.* (20):133-155. doi:https://doi.org/10.23962/10539/23572
8. Bada M, Sasse AGBM, Nurse JRC. Cyber Security Awareness Campaigns: Why do they fail to change behaviour? In: *Journal of the International Conference on Cyber Security for Sustainable Society 2015.* Vol abs/1901.02672. ; 2015.
9. Sasse A. Scaring and bullying people into security won't work. Lesk M, MacKie-Mason J, eds. *IEEE Security & Privacy.* 2015;13(3):80-83.
10. Thycotic. Cyber Security Teams Survey Report 2019 | IT Security UK & Germany. Thycotic. https://thycotic.com/resources/cyber-security-executives-survey-report-europe/. Accessed March 4, 2019.
11. Das S, Kim TH-J, Dabbish LA, Hong JI. The Effect of Social Influence on Security Sensitivity. In: *Proceedings of the Tenth USENIX Conference on Usable Privacy and Security.* SOUPS'14. Berkeley, CA, USA: USENIX Association; 2014:143–157. http://dl.acm.org/citation.cfm?id=3235838.3235851.
12. Burns S, Roberts L. Applying the Theory of Planned Behaviour to predicting online safety behaviour. *Crime Prevention & Community Safety.* 2013;15:48-64. doi:10.1057/cpcs.2012.13
13. Lewis K, Kaufman J, Christakis N. The Taste for Privacy: An Analysis of College Student Privacy Settings in an Online Social Network. *Journal of Computer-Mediated Communication.* 2008;14(1):79-100. doi:10.1111/j.1083-6101.2008.01432.x
14. Poller A, Kocksch L, Türpe S, Epp FA, Kinder-Kurlanda K. Can Security Become a Routine? A Study of Organizatoinal Change in an Agile Software Development Group. In: Portland, OR, USA; 2017.
15. CPNI. Workmate Bingo. Workplace behaviours campaign. https://www.cpni.gov.uk/workplace-behaviours. Published 2015. Accessed March 8, 2019.

16. Computer Glossary, Computer Terms - Technology Definitions and Cheat Sheets from WhatIs.com - The Tech Dictionary and IT Encyclopedia. https://whatis.techtarget.com/. Accessed January 28, 2019.
17. BSI. Glossary of cyber security terms. https://www.bsigroup.com/en-GB/Cyber-Security/Cyber-security-for-SMEs/Glossary-of-cyber-security-terms/.
18. NICCS. Glossary. https://niccs.us-cert.gov/about-niccs/glossary.
19. ENISA. ENISA overview of cybersecurity and related terminology. 2017. https://www.enisa.europa.eu/publications/enisa-position-papers-and-opinions/enisa-overview-of-cybersecurity-and-related-terminology.
20. Hughes B. The Bishop Fox Cybersecurity Style Guide. 2019. https://www.bishopfox.com/blog/2018/02/hello-world-introducing-the-bishop-fox-cybersecurity-style-guide/.
21. Microsoft writing style guide. https://docs.microsoft.com/en-us/style-guide/welcome/. Published 2018.
22. ENISA. A good practice guide of using taxonomies in incident prevention and detection. https://www.enisa.europa.eu/publications/using-taxonomies-in-incident-prevention-detection. Published 2017. Accessed January 15, 2019.
23. Craigen D, Diakun-Thibault N, Purse R. Defining Cybersecurity. *Technology Innovation Management Review*. 2014;4:13-21. doi:http://doi.org/10.22215/timreview/835
24. Thomas J, McDonagh D. Shared language: Towards more effective communication. *The Australasian medical journal*. 2013;6(1):46-54. doi:10.4066/AMJ.2013.1596
25. van Deursen N. HI-Risk: a Socio-Technical Method for the Identification and Monitoring of Healthcare Information Security Risks in the Information Society. 2014.
26. Avgerinou MD, Pettersson R. Toward a Cohesive Theory of Visual Literacy. *Journal of Visual Literacy*. 2011;30(2):1-19. doi:10.1080/23796529.2011.11674687
27. Common Core: Paths to 21st-Century Success. Center for Teaching Quality #CTQCollab. https://www.teachingquality.org/coresuccess/. Published June 12, 2014. Accessed March 8, 2019.
28. Marabella A. Communication Theories: an infographics development project. 2014.
29. Kędra J. What does it mean to be visually literate? Examination of visual literacy definitions in a context of higher education. *Journal of Visual Literacy*. 2018;37(2):67-84. doi:10.1080/1051144X.2018.1492234
30. Avgerinou M. Towards a visual literacy index. *Journal of Visual Literacy*. 2007;27(1).
31. Brown S. *The Doodle Revolution*. New York: Portfolio/Penguin; 2014.
32. The Future of Jobs Report 2018. World Economic Forum. https://www.weforum.org/reports/the-future-of-jobs-report-2018/. Accessed February 4, 2019.
33. The Hague Security Delta. Wanted Security Professionals. An analysis of job advertisements. https://www.thehaguesecuritydelta.com/media/com_hsd/report/193/document/Wanted-Security-Professionals-An-Analysis-of-Job-Advertisements.pdf. Published 2018. Accessed February 4, 2019.
34. Kharb P, Samanta PP, Jindal M, Singh V. The learning styles and the preferred teaching-learning strategies of first year medical students. *J Clin Diagn Res*. 2013;7(6):1089-1092. doi:10.7860/JCDR/2013/5809.3090
35. F. Sharevski, A. Trowbridge, J. Westbrook. Novel approach for cybersecurity workforce development: A course in secure design. In: *2018 IEEE Integrated STEM Education Conference (ISEC)*. ; 2018:175-180. doi:10.1109/ISECon.2018.8340471

36. Hasso Plattner Institute of Design, d.school Stanford. Virtual crash course in Design Thinking. https://dschool.stanford.edu/resources-collections/a-virtual-crash-course-in-design-thinking. Accessed January 15, 2019.

37. LeBlanc C. *Tackling Wicked Problems. An OER for Students at PSU.* Plymouth State University; 2017. https://wicked-problem.press.plymouth.edu/front-matter/about/.

38. Jacobson J, Turner-Rahman G. *Visual Intelligence: Bridging the Gap from Visual Literacy to Visual Reasoning.*; 2019.

39. Pemberton Levy H. Lessons in How to Implement People-Centric Security. Gartner. https://www.gartner.com/smarterwithgartner/lessons-in-how-to-implement-people-centric-security/. Published 2015. Accessed February 19, 2019.

40. Bloom BS, Krathwohl DR. *Taxonomy of Educational Objectives: The Classification of Educational Goals.* New York, NY, USA; 1956.

41. Anderson LW, Krathwohl DR, eds. *A Taxonomy for Learning, Teaching, and Assessing: A Revision of Bloom's Taxonomy of Educational Objectives.* Boston, MA: Allyn & Bacon; 2001.

42. Anttila J, Savola R, Kajava J, Lindfors J, Röning J. Fulfilling the Needs for Information Security Awareness and Learning in Information Society. In: *Proceedings of the 6th Annual Security Conference.* Las Vegas; 2007. http://www.isy.vcu.edu/~gdhillon/Old2/secconf/secconf07/PDFs/21.pdf.

43. van Niekerk J, von Solms R. Using Bloom's Taxonomy for Information Security Education. In: *8th World Conference on Information Security Education (WISE).* Vol AICT-406. Bento Gonçalves, Brazil: Springer; 2009:280-287. https://hal.inria.fr/hal-01463654.

44. Harris MA, Patten KP. Using Bloom's and Webb's Taxonomies to Integrate Emerging Cybersecurity Topics into a Computing Curriculum. *Journal of Information Systems Education.* 2015;26(3):219-234.

45. Howard-Jones PA. Neuroscience and education: Myths and messages. *Nature Reviews Neuroscience.* 2014;(15):817-824. doi:doi:10. 1038/nrn3817

46. Fleming ND. I'm different; not dumb. Modes of presentation (V.A.R.K.) in the tertiary classroom. In: Zelmer A, ed. *Proceedings of the 1995 Annual Conference of the Higher Education and Research Development Society of Australasia (HERDSA).* Vol 18. ; 1995. http://www.vark-learn.com/wp-content/uploads/2014/08/different_not_dumb.pdf.

47. Introduction to VARK | VARK. http://vark-learn.com/introduction-to-vark/. Accessed March 18, 2019.

48. Fleming ND. What Is the Link between VARK and Communication? https://www.youtube.com/watch?v=fm_1Cj6r9DI. Published 2011. Accessed March 18, 2019.

49. Pattinson M, Butavicius M, Parsons K, Mccormac A, Calic D. Managing information security awareness at an Australian bank: a comparative study. *Information & Computer Security.* 2017;25(2):181-189. doi:10.1108/ICS-03-2017-0017

50. Kirova D, Baumöl U. Factors that Affect the Success of Security Education, Training, and Awareness Programs: A Literature Review. *JITTA : Journal of Information Technology Theory and Application.* 2018;19(4):56-82.

51. Staalduine J van. ABN Amro "fopt" personeel met nep-phishingmail over terugkeer kerstpakket. de Volkskrant. https://www.volkskrant.nl/gs-b4daf7d0. Published November 24, 2017. Accessed November 2, 2019.

52. Diaz A, Sherman AT, Joshi A. Phishing in an Academic Community: A Study of User Susceptibility and Behavior. 2018.

53. J. G. Mohebzada, A. E. Zarka, A. H. Bhojani, A. Darwish. Phishing in a university community: Two large scale phishing experiments. In: *2012 International Conference on Innovations in Information Technology (IIT).* ; 2012:249-254. doi:10.1109/INNOVATIONS.2012.6207742

54. Tschakert KF, Ngamsuriyaroj S. Effectiveness of and user preferences for security awareness training methodologies. *Heliyon.* 2019;5(6):e02010. doi:10.1016/j.heliyon.2019.e02010

55. D. D. Caputo, S. L. Pfleeger, J. D. Freeman, M. E. Johnson. Going Spear Phishing: Exploring Embedded Training and Awareness. *IEEE Security & Privacy.* 2014;12(1):28-38. doi:10.1109/MSP.2013.106

56. Gordon WJ, Wright A, Glynn RJ, et al. Evaluation of a mandatory phishing training program for high-risk employees at a US healthcare system. *Journal of the American Medical Informatics Association.* 2019;26(6):547-552. doi:10.1093/jamia/ocz005

57. Reijnders E. *Basisboek Interne Communicatie.* 7th ed. Assen, the Netherlands: van Gorcum; 2010.

58. Assal H, Chiasson S. Security in the Software Development Lifecycle. In: *Fourteenth Symposium on Usable Privacy and Security (SOUPS 2018).* Baltimore, MD: USENIX Association; 2018:281–296. https://www.usenix.org/conference/soups2018/presentation/assal.

59. Mintzberg H. *Mintzberg on Management.* New York, NY, USA: The Free Press; 1989.

60. Schlienger T, Teufel S. Information Security Culture: The socio-cultural dimension in information security management. In: Ghonaimy MA, El-Hadidi MT, Aslan HK, eds. *Security in the Information Society.* Vol 86. IFIP Advances in Information and Communication Technology. Boston, MA: Springer; 2002:191-201.

61. Cyber Security Culture in organisations — ENISA. https://www.enisa.europa.eu/publications/cyber-security-culture-in-organisations. Accessed March 7, 2019.

62. Martins A, Elofe J. Information security culture. In: Ghonaimy MA, El-Hadidi MT, Aslan HK, eds. *Security in the Information Society.* Vol 86. IFIP Advances in Information and Communication Technology. Boston, MA: Springer; 2002.

63. van Niekerk J, von Solms R. A holistic framework for the fostering of an information security sub-culture in organizations. In: *Proceedings of the ISSA 2005.* South Africa; 2005:1-13.

64. Stahl BC, Doherty NF, Shaw M. Information security policies in the UK healthcare sector: a critical evaluation. *Information Systems Journal.* 2012;22(1):77-94.

65. Guo KH, Yuan Y. The effects of multilevel sanctions on information security violations: A mediating model. *Information & Management.* 2012;49(6):320-326.

66. Coles-Kemp L, Theoharidou M. Insider Threat and Information Security Management. In: Probst C, Hunker J, Gollmann D, Bishop M, eds. *Insider Threats in Cyber Security.* Vol 49. Advances in Information Security. Springer, Boston, MA; 2010.

67. Srivastava SB, Goldberg A. Language as a Window into Culture. *California Management Review.* 2017;60(1):56-69. doi:10.1177/0008125617731781

68. Gold S. Securing the National Health Service. *Computer Fraud & Security.* 2010;(5):11-14.

69. Renaud K, Goucher W. Health service employees and information security policies: an uneasy partnership? *Information Management and Computer Security.* 2012;20(4):296-311.

70. Quirke B. *Making the Connections: Using Internal Communication to Turn Strategy into Action.* 2nd edition. London and New York: Routledge; 2017.

71. Friedman AL, Miles S. *Stakeholders. Theory and Practice.* Oxford University Press; 2006.

72. Freeman RE. What is stakeholder theory? https://www.youtube.com/watch?v=bIRUaLcvPe8&feature=relmfu&nomobile=1. Accessed January 8, 2019.

73. Ramírez R. Stakeholder analysis and conflict management. In: *Cultivating Peace: Conflict and Collaboration in Natural Resource Management.* Ottawa, Canada: International Development Research Centre; 1999. https://idl-bnc-idrc.dspacedirect.org/bitstream/handle/10625/29303/IDL-29303.pdf?sequence=1. Accessed January 8, 2019.

74. d'Herbemont BC, César B. *Managing Sensitive Projects: A Lateral Approach*. New York: Routledge; 1998.

75. Bourne L. *Making Projects Work: Effective Stakeholder and Communication Management*. CRC Press; 2015.

76. Petronio S. *Boundaries of Privacy. Dialectics of Disclosure*. New York: SUNY Press; 2002.

77. Miller KW, Voas J, Hurlburt GF. BYOD: Security and Privacy Considerations. *IT Professional*. 2012;14(September/October 2012):53-55. doi:10.1109/MITP.2012.93

78. Chang SE, Liu AY, Lin S. Exploring privacy and trust for employee monitoring. *Industrial Management & Data Systems*. 2015;115(1):88-106.

79. Eastin MS, Brinson NH, Doorey A, Wilcox G. Living in a big data world: Predicting mobile commerce activity through privacy concerns. *Computers in Human Behavior*. 2016;58:214-220. doi:https://doi.org/10.1016/j.chb.2015.12.050

80. Kotler P, Kartajaya H, Setiawan I. *Marketing 4.0: Moving from Traditional to Digital*. Hoboken, New Jersey: John Wiley & Sons; 2017.

81. Herath T, Rao HR. Encouraging information security behaviors in organizations: Role of penalties, pressures and perceived effectiveness. *Decision Support Systems*. 2009;47(2):154-165.

82. Johnson ME. Managing Information Risk and the Economics of Security. In: Johnson ME, ed. *Managing Information Risk and the Economics of Security*. Boston, MA: Springer.

83. Scholl F. Better security through storytelling. CSO Online. https://www.csoonline.com/article/3162924/better-security-through-storytelling.html. Published January 30, 2017. Accessed March 7, 2019.

84. Quigley K, Burns C, Stallard K. 'Cyber Gurus': A rhetorical analysis of the language of cybersecurity specialists and the implications for security policy and critical infrastructure protection. *Government Information Quarterly*. 2015;32(2):108-117. doi:10.1016/j.giq.2015.02.001

85. S. T. Lawson, S. K. Yeo, Haoran Yu, E. Greene. The cyber-doom effect: The impact of fear appeals in the US cyber security debate. In: *2016 8th International Conference on Cyber Conflict (CyCon)*. ; 2016:65-80. doi:10.1109/CYCON.2016.7529427

86. Bruijn H de, Janssen M. Building Cybersecurity Awareness: The need for evidence-based framing strategies. *Government Information Quarterly*. 2017;34(1):1-7. doi:https://doi.org/10.1016/j.giq.2017.02.007

87. Ministerie van Defensie. *Persconferentie: MIVD Verstoorde GRU Cyberoperatie in Den Haag*. https://www.youtube.com/watch?v=jMBdN1n2uBI&feature=youtu.be. Accessed March 18, 2019.

88. Kaspersky Lab Benelux. *This Is a Message to the Dutch Government*.; 2018. https://www.youtube.com/watch?v=rc8r3UFaXaM. Accessed March 9, 2019.

89. Qing T, Ng B-Y, Kankanhalli A. Individual's Response to Security Messages: A Decision-Making Perspective. In: *Decision Support for Global Enterprises*. Vol 2. Annals of information systems. ; 2007:177-191. doi:10.1007/978-0-387-48137-1_10

90. Puhakainen P, Siponen M. Improving Employees' Compliance Through Information Systems Security Training: An Action Research Study. *MIS Quarterly*. 2010;34(4):757-778. doi:10.2307/25750704

91. Vishwanath A, Herath T, Chen R, Wang J, Rao HR. Why do people get phished? Testing individual differences in phishing vulnerability within an integrated, information processing model. *Decision Support Systems*. 2011;51(3):576-586. doi:10.1016/j.dss.2011.03.002

92. Hallam C, Zanella G. Online self-disclosure: The privacy paradox explained as a temporally discounted balance between concerns and rewards. *Computers in Human Behavior.* 2017;68:217-227. doi:https://doi.org/10.1016/j.chb.2016.11.033

93. Barth S, de Jong MDT. The privacy paradox: Investigating discrepancies between expressed privacy concerns and actual online behavior - A systematic literature review. *Telematics and informatics.* 2017;34(7):1038–1058. doi:10.1016/j.tele.2017.04.013

94. Ghosh I, Singh V. Using cognitive dissonance theory to understand privacy behavior. *Proceedings of the Association for Information Science and Technology.* 2017;54(1):679-681. doi:10.1002/pra2.2017.14505401114

95. LastPass. The password paradox and why our personalities will get us hacked. http://prod.cdata.app.sprinklr.com/DAM/434/LastPass_ExecutiveSummary_fina-88e8a5a2-00cb-4a09-b363-e01a45f829d6-1389898992.pdf. Published 2016.

96. Thaler RH, Sunstein CR. *Nudge. Improving Decisions about Health, Wealth, and Happiness.* Yale University Press; 2008.

97. Renaud K, Zimmermann V. Ethical guidelines for nudging in information security & privacy. *International Journal of Human-Computer Studies.* 2018;120:22-35. doi:10.1016/j.ijhcs.2018.05.011

98. Sunstein CR, Reisch LA. *Trusting Nudges: Toward a Bill of Rights for Nudging.* 1st edition. Oxon and New York; 2019.

99. Lai Y-L, Hui K-L. Internet opt-in and opt-out: investigating the roles of frames, defaults and privacy concerns. In: *Proceedings of the 2006 ACM SIGMIS CPR Conference on Computer Personnel Research: Forty Four Years of Computer Personnel Research: Achievements, Challenges & the Future.* Claremont, California, USA: ACM; 2006:253-263.

100. van Bavel R, Rodríguez-Priego N. *Nudging Online Security Behaviour with Warning Messages: Results from an Online Experiment.* Luxembourg: European Uninion; 2016.

101. Turland J, Coventry L, Jeske D, Briggs P, van Moorsel A. *Nudging Towards Security: Developing an Application for Wireless Network Selection for Android Phones.*; 2015. doi:10.1145/2783446.2783588

102. Ajzen I. The theory of planned behavior. *Organizational Behavior and Human Decision Processes.* 1991;50(2):179-211. doi:https://doi.org/10.1016/0749-5978(91)90020-T

103. Wagemans J, Elder JH, Kubovy M, et al. A century of Gestalt psychology in visual perception: I Perceptual grouping and figure–ground organization. *Psychological Bulletin.* 2012;138(6):1172-1217. doi:10.1037/a0029333

104. J. Garae, R. K. L. Ko, S. Chaisiri. UVisP: User-centric Visualization of Data Provenance with Gestalt Principles. In: *2016 IEEE Trustcom/BigDataSE/ISPA.* ; 2016:1923-1930. doi:10.1109/TrustCom.2016.0294

105. Köppen M. Gestalt Aspects of Security Patterns. In: Sako H, Franke KY, Saitoh S, eds. *Computational Forensics.* Springer Berlin Heidelberg; 2011:1-12.

106. Anderson BB, Jenkins JL, Vance A, Kirwan CB, Eargle D. Your memory is working against you: How eye tracking and memory explain habituation to security warnings. *Decision Support Systems.* 2016;92:3-13. doi:https://doi.org/10.1016/j.dss.2016.09.010

107. Vance A, Jenkings JL, Anderson BB. Tuning out security warnings: a longitudinal examination of habituation through fMRY, eye tracking, and field experiments. *MIS Quarterly.* 2018;42(2):355-380.

108. Kolb N, Bartsch S, Volkamer M, Vogt J. Capturing Attention for Warnings about Insecure Password Fields – Systematic Development of a Passive Security Intervention. In: Tryfonas T, Askoxylakis I, eds. *Human Aspects of Information Security, Privacy, and Trust.* Springer International Publishing; 2014:172-182.

109. Porter Felt A, Ainslie A, Reeder RW, et al. Improving SSL Warnings: Comprehansion and Adherence. In: *Proceedings of the Conference Human Factors and Computing Systems*. Seoul; 2015.

110. Stanford Law School & d.school. The Legal Design Lab. https://law.stanford.edu/organizations/pages/legal-design-lab/. Accessed January 25, 2019.

111. Legal Geek. https://www.legalgeek.co/LegalDesign/.

112. University of Luxembourg. Legal Informatics Luxembourg. http://www.luxli.lu/legal-design/.

113. Legal Design Alliance. https://www.legaldesignalliance.org.

114. Rossi A, Palmirani M. From words to images through legal visualization. In: *AICOL VI-X*. ; 2018:72-85.

115. Zarkada A. *Concepts and Constructs for Personal Branding: An Exploratory Literature Review Approach*.; 2012. doi:10.2139/ssrn.1994522

116. Vallas SP, Cummins ER. Personal Branding and Identity Norms in the Popular Business Press: Enterprise Culture in an Age of Precarity. *Organization Studies*. 2015;36(3):293-319. doi:10.1177/0170840614563741

117. Rangarajan D, Gelb BD, Vandaveer A. Strategic personal branding—And how it pays off. *Business Horizons*. 2017;60(5):657-666. doi:https://doi.org/10.1016/j.bushor.2017.05.009

118. Frankland J. *INSecurity*. Rethink Press; 2017.

119. IEEE Symposium on visualization for cyber security. Vizsec. https://vizsec.org/. Accessed March 15, 2019.

120. Graphical Models for Security 2019. https://gramsec.uni.lu/. Accessed November 6, 2019.

121. Marty R. *Applied Security Visualization*. Addison-Wesley; 2008.

122. Liebow-Feeser J. Randomness 101: LavaRand in Production. The Cloudflare Blog. https://blog.cloudflare.com/randomness-101-lavarand-in-production/. Published November 6, 2017. Accessed November 24, 2019.

123. Assal H, Chiasson S, Biddle R. Cesar: Visual representation of source code vulnerabilities. In: *2016 IEEE Symposium on Visualization for Cyber Security (VizSec)*. ; 2016:1-8. doi:10.1109/VIZSEC.2016.7739576

124. Comodo Threat Research Lab. Comodo Threat Research Labs Q3 2017 REPORT. https://www.comodo.com/ctrlquarterlyreport/q3/Comodo_Q3Report_111417_HR.pdf. Accessed October 5, 2019.

125. Arendt D, Best D, Burtner R, Paul CL. CyberPetri at CDX 2016: Real-time network situation awareness. In: ; 2016:1-4. doi:10.1109/VIZSEC.2016.7739584

126. Post T, Wischgoll T, Bryant AR, Hamann B, Müller P, Hagen H. Visually guided flow tracking in software-defined networking. In: ; 2016:1-4. doi:a

127. den Braber F, Brændeland G, Dahl HEI, et al. *The CORAS Model-Based Method for Security Risk Analysis*. Oslo: SINTEF; 2006. https://www.uio.no/studier/emner/matnat/ifi/INF5150/h06/undervisningsmateriale/060930.CORAS-handbook-v1.0.pdf.

128. A. Sen, S. Madria. Risk Assessment in a Sensor Cloud Framework Using Attack Graphs. *IEEE Transactions on Services Computing*. 2017;10(6):942-955. doi:10.1109/TSC.2016.2544307

129. L. Gallon, J. Bascou. CVSS Attack Graphs. In: *2011 Seventh International Conference on Signal Image Technology & Internet-Based Systems*. ; 2011:24-31. doi:se

130. Applebaum A. Visualizing ATT&CK. Medium. https://medium.com/mitre-attack/visualizing-attack-f5e1766b42a6. Published March 4, 2019. Accessed November 2, 2019.

131. Graph visualization use cases: cyber security. Cambridge Intelligence. https://cambridge-intelligence.com/use-cases-graph-visualization-cyber-security/. Published September 25, 2017. Accessed November 2, 2019.

132. Noel S, Harley E, Tam KH, Limiero M, Share M. Chapter 4 - CyGraph: Graph-Based Analytics and Visualization for Cybersecurity. In: Gudivada VN, Raghavan VV, Govindaraju V, Rao CR, eds. *Handbook of Statistics*. Vol 35. Elsevier; 2016:117-167. doi:10.1016/bs.host.2016.07.001

133. OWASP SAMM Project - OWASP. Owasp.org. https://www.owasp.org/index.php/OWASP_SAMM_Project. Accessed October 30, 2019.

134. Li Y, Huang G, Wang C, Li Y. Analysis framework of network security situational awareness and comparison of implementation methods. *EURASIP Journal on Wireless Communications and Networking*. 2019;2019(1):205. doi:10.1186/s13638-019-1506-1

135. Albanese M, Cam H, Jajodia S. Automated Cyber Situation Awareness Tools and Models For Improving Analyst Performance. *Advances in Information Security*. 2014;61:47-60. doi:10.1007/978-3-319-10374-7_3

136. *Financial Institutions: How to Protect Customers from Advanced Malware in 2016*. Wontok SafeCentral; 2016. https://wontok.com/wp-content/uploads/2018/07/Wontok-HowToProtectYourBankCustomersIn2016_v2r4.pdf.

137. TREsPASS. Visualising virtual infrastructure. https://visualisation.trespass-project.eu/?p=293. Published 2016. Accessed March 17, 2019.

138. Spitzner L. Security Awareness Maturity Model / Kit. SANS Security Awareness. https://www.sans.org//security-awareness-training/blog/security-awareness-maturity-model-kit. Published 2018. Accessed March 20, 2019.

139. F-secure. Ransomware Families. https://heimdalsecurity.com/blog/wp-content/uploads/ransowmare-families-f-secure-1.jpg. Accessed November 3, 2019.

140. ENISA. European Cyber Security Month Roadmap. https://www.enisa.europa.eu/topics/cybersecurity-education/european-cyber-security-month/2013/european-cyber-security-month-roadmap/view. Accessed March 11, 2019.

141. Wazir F. Can NL trust 5G? A conceptual model for cyber security supervision of 5G in the Netherlands. https://www.csacademy.nl/scripties/februari-2019/129-can-nl-trust-5g.

142. McGilvray D, Price J, Redman T. fishbone-diagram. Dataleaders.com. https://dataleaders.files.wordpress.com/2017/05/fishbone-diagram.png. Accessed March 25, 2019.

143. Evans N, Price J. Barriers to the effective deployment of information assets: an executive management perspective. *Interdisciplinary Journal of Information, Knowledge, and Management*. 2012;7.

144. Protecting Yourself from Bad Rabbit Ransomware - Security News - Trend Micro PH. https://www.trendmicro.com/vinfo/ph/security/news/cyber-attacks/protecting-yourself-from-bad-rabbit-ransomware. Accessed March 20, 2019.

145. Subway map to agile practices. *Agile Alliance*. https://www.agilealliance.org/agile101/subway-map-to-agile-practices/. Accessed October 19, 2019.

146. Cybersecurity Career Pathway. CyberSeek Project and partners Burning Glass Technologies, CompTIA, National Initiative for Cybersecurity Education (NICE). https://www.cyberseek.org/pathway.html. Accessed March 20, 2019.

147. World Economic Forum. Global Risks 2018 Interconnections Map. Reports. http://reports.weforum.org/global-risks-2018/global-risks-landscape-2018/#risks. Accessed March 17, 2019.

148. Infographics at the NCSC. National Cyber Security Centre. https://www.ncsc.gov.uk/information/infographics-ncsc. Published 2018. Accessed March 20, 2019.

149. McCanless D, Evans T, Barton P, Tomasevic S. World's Biggest Data Breaches & Hacks. Information is Beautiful. https://informationisbeautiful.net/visualizations/worlds-biggest-data-breaches-hacks/. Accessed March 27, 2019.

150. McKenna S, Staheli D, Fulcher C, Meyer M. BubbleNet: A Cyber Security Dashboard for Visualizing Patterns. *Computer Graphics Forum*. 2016;35(3):281-290. doi:10.1111/cgf.12904

151. CISO Platform. Understanding difference between Cyber Security & Information Security. https://www.cisoplatform.com/profiles/blogs/understanding-difference-between-cyber-security-information. Published July 22, 2016. Accessed November 2, 2019.

152. Tokody D, Albini A, Ady L, Rajnai Z, Pongrácz F. Safety and Security through the Design of Autonomous Intelligent Vehicle Systems and Intelligent Infrastructure in the Smart City. 2018;16:384-396. doi:10.7906/indecs.16.3.11

153. The Periodic Table of Security | IFSEC Global. *IFSEC Global | Security and Fire News and Resources*. December 2018. https://www.ifsecglobal.com/security/periodic-table-of-security/. Accessed March 20, 2019.

154. S. Peryt, J. Andre Morales, W. Casey, A. Volkmann, B. Mishra, Y. Cai. Visualizing a Malware Distribution Network. In: *2016 IEEE Symposium on Visualization for Cyber Security (VizSec).*; 2016:1-4. doi:10.1109/VIZSEC.2016.7739585

155. Walton S, Maguire E, Chen M. Multiple queries with conditional attributes (QCATs) for anomaly detection and visualization. In: *Proceedings of the Eleventh Workshop on Visualization for Cyber Security*. Paris, France: ACM; 2014:17-24.

156. *Global Cybersecurity Index (GCI) 2017*. International Telecommunication Union (ITU); 2017. https://www.itu.int/dms_pub/itu-d/opb/str/D-STR-GCI.01-2017-R1-PDF-E.pdf.

157. Kaspersky. Kaspersky Cyberthreat real-time map. MAP | Kaspersky Cyberthreat real-time map. https://cybermap.kaspersky.com/. Accessed November 2, 2019.

158. Check Point. Live Cyber Threat Map. https://threatmap.checkpoint.com/. Accessed November 2, 2019.

159. FireEye. Cyber Threat Map. FireEye. https://www.fireeye.com/cyber-map/threat-map.html. Accessed November 2, 2019.

160. Akamai. Real-time web monitor. https://www.akamai.com/us/en/resources/visualizing-akamai/real-time-web-monitor.jsp. Accessed November 2, 2019.

161. Fortinet Threat Map. http://threatmap.fortiguard.com/. Accessed November 2, 2019.

162. Amado R. Ransomware-as-a-service: The Business Case. Digital Shadows. https://www.digitalshadows.com/blog-and-research/ransomware-as-a-service-the-business-case/. Published November 22, 2016. Accessed November 2, 2019.

163. Calorio P. Record of processing actitivities, data protection officer and privacy impact assessment: the "crossings". Twitter. https://twitter.com/PietroCalorio/status/967411827259969536. Accessed November 6, 2019.

164. Bates C, Clark T, Davis T, Fisch N. Technology in Higher Education: Information Security Leadership. 2016. https://library.educause.edu/resources/2016/3/technology-in-higher-education-information-security-leadership.

165. Deursen N van. Wie is de meest gezochte informatiebeveiliger? *IB Magazine*. 2018;(#4):12-17.

166. Barendse J, Li E. Visualising attack graphs for DBIR 2016. TREsPASS project. https://visualisation.trespass-project.eu/?p=275. Accessed March 25, 2019.

167. Cyber Security Horoscope 2019 | LinkedIn. https://www.linkedin.com/pulse/cyber-security-horoscope-2019-nicole-van-deursen-phd/. Accessed October 31, 2019.

168. Occupations Radar for Safety & Security. Security Talent. https://securitytalent.nl/news/occupations-radar-for-safety-security. Published 2018. Accessed March 20, 2019.

169. Frye E. The Twilight Zone of Cyber Response. MITRE Cybersecurity. https://www.mitre.org/capabilities/cybersecurity/overview/cybersecurity-blog/the-twilight-zone-of-cyber-response. Published December 10, 2013. Accessed March 20, 2019.

170. Mobile Security iOS App. Evan Prowten. http://www.evanprowten.com/mobile-security-ios-app. Accessed March 21, 2019.

171. Design Workshop for EU General Data Protection Regulation. *Legal Design Lab*. July 2017. http://www.legaltechdesign.com/design-workshop-for-eu-general-data-protection-regulation/. Accessed November 3, 2019.

172. *Cybersecurity Awareness - Phishing Attacks - from SANS.Org*. https://www.youtube.com/watch?v=5RHeJAEdiEc. Accessed November 3, 2019.

173. Infosecinstitute. *An Introduction to Cybersecurity Careers*. https://www.youtube.com/watch?v=-AkuKKJ8dN0. Accessed November 3, 2019.

174. Muenchow O. *Security Awareness Video: 7 Tips for Your Employees to Be Able to Identify and Avoid Risks*. https://www.youtube.com/watch?v=i0iLy8racHI. Accessed November 3, 2019.

175. ISACA. *Overview of Digital Forensics*. https://www.youtube.com/watch?v=ZUqzcQc_syE. Accessed November 3, 2019.

176. CIAS. CIAS Cyber Threat Defender. http://cias.utsa.edu/ctd_cards.php. Accessed March 25, 2019.

177. Reflections and Graphic Recording from SWIFT Innotribe at Sibos 2018. Collective Next. https://collectivenext.com/blog/swift-innotribe-sibos-2018-financial-services-conference-graphic-recording-facilitation-program-curation. Published November 27, 2018. Accessed October 20, 2019.

178. TREsPASS. Paper prototyping: Cloud case study. https://visualisation.trespass-project.eu/?p=286. Accessed March 25, 2019.

179. Microsoft, Frost & Sullivan. Cybersecurity threats to cost organisations in Singapore US$17.7 billion in economic losses. Microsoft Singapore News Center. https://news.microsoft.com/en-sg/2018/05/18/cybersecurity-threats-to-cost-organisations-in-singapore-us17-7-billion-in-economic-losses/. Published May 18, 2018. Accessed March 12, 2019.

180. Brand D. Ten Cybersecurity Action Items for CAEs and Internal Audit Departments. KnowledgeLeader provided by Protiviti. https://info.knowledgeleader.com/ten-cybersecurity-action-items-for-caes-and-internal-audit-departments-old. Accessed November 2, 2019.

181. Chan Z. Cybersecurity and you: the iceberg effect. *HMW Singapore*. July 2018. https://www.pressreader.com/singapore/hwm-singapore/20180701/281887299056714. Accessed November 2, 2019.

182. Deloitte US. Diving deeper into federal cybersecurity attacks. Deloitte United States. https://www2.deloitte.com/us/en/pages/public-sector/articles/federal-cybersecurity-risk-management.html. Accessed November 2, 2019.

183. Maor E. Cybercrime Ecosystem: Everything Is for Sale. Security Intelligence. https://securityintelligence.com/cybercrime-ecosystem-everything-is-for-sale/. Published June 15, 2015. Accessed March 27, 2019.

184. Aiuken Cybersecurity. Security Operation Center. https://www.aiuken.com/en/services/security-operation-center. Accessed November 2, 2019.

185. Evans B. If Cyber Security was a Dagwood sandwich, Identity and Access Management would be the bread. Direct2Dell. https://blog.dell.com/en-us/if-cyber-security-was-a-dagwood-sandwich-identity-and-access-management-would-be-the-bread/. Published September 13, 2013. Accessed November 2, 2019.

186. Zachary J. Notes on Cybersecurity Psychology. John Zachary. http://johnzachary.com/blog/psychology-of-cybersecurity/. Published December 5, 2016. Accessed November 2, 2019.

187. European Parliament. The top cyber threats. https://www.europarl.europa.eu/news/en/headlines/security/20160701STO34371/cyber-security-new-rules-to-protect-europe-s-infrastructure. Published July 5, 2016. Accessed November 3, 2019.

188. Lowe J, Duffy S. Security Sessions | Is Your C-Suite Prepared for the Cyber Future? Electric Energy Online. http://electricenergyonline.com/energy/magazine/1028/article/Security-Sessions-Is-Your-C-Suite-Prepared-for-the-Cyber-Future-.htm. Accessed March 20, 2019.

189. Neuways, McRea J. Time to Improve Your Cyber Security - Neuways Blog. *Neuways.* February 2019. https://www.neuways.com/neuways-blog/time-to-improve-your-cyber-security/. Accessed March 27, 2019.

190. Taylor A. Data Science Hunting Funnel. Austin Taylor. http://www.austintaylor.io/network/traffic/threat/data/science/hunting/funnel/machine/learning/domain/expertise/2017/07/11/data-science-hunting-funnel/. Accessed March 27, 2019.

191. Privacy Icons Forum. https://www.privacyiconsforum.eu/. Accessed November 3, 2019.

192. Holtz L-E, Zwingelberg H, Hansen M. Privacy Policy Icons. In: Camenisch J, Fischer-Hübner S, Rannenberg K, eds. *Privacy and Identity Management for Life.* Berlin, Heidelberg: Springer Berlin Heidelberg; 2011:279-285. doi:10.1007/978-3-642-20317-6_15

193. Privacy icons - Privacy Patterns. https://privacypatterns.org/patterns/Privacy-icons. Accessed February 7, 2019.

194. Zone Alarm. Online threats 101. https://i.pinimg.com/1200x/65/30/c2/6530c2468d18c1a9664e5db05d0108e9.jpg. Accessed November 3, 2019.

195. Stretch J. Common Ports Cheat Sheet from Cheatography. Commonly used TCP / UDP port numbers.... *LastStepPin.* June 2019. https://www.laststeppin.com/share/pin-it/25589/2019/. Accessed November 3, 2019.

196. Gunn J. Social Engineering and How to Win the Battle for Trust [Infographic]. OneSpan. https://www.onespan.com/blog/social-engineering-win-battle-trust-infographic. Accessed March 27, 2019.

196b. Global Cybersecurity Index (GCI) 2017. International Telecommunication Union (ITU); 2017. Available from: https://www.itu.int/dms_pub/itu-d/opb/str/D-STR-GCI.01-2017-R1-PDF-E.pdf

197. Krum R. *Cool Infographics. Effective Communication with Data Visualization and Design.* Indianapolis: John Wiley & Sons; 2014.

198. Dit zijn geen infographics (echt niet). Lakmoes - kenniscommunicatie. https://www.studio-lakmoes.nl/blog/dit-zijn-geen-infographics. Accessed March 14, 2019.

199. VanDam L. Introducing The Psychology of Passwords. The LastPass Blog. https://blog.lastpass.com/2016/09/infographic-introducing-the-psychology-of-passwords.html/. Published September 28, 2016. Accessed March 9, 2019.

200. Groenland T. Hackers Handshake - portraits by Tobias Groenland. Tobias Groenland. https://www.tobiasgroenland.nl/hackers-handshake/. Accessed November 8, 2019.

201. Nationaal Cyber Security Centrum. *Leidraad Coordinated Vulnerability Disclosure.*; 2018. https://www.ncsc.nl/documenten/publicaties/2019/mei/01/cvd-leidraad. Accessed November 20, 2019.

202. van 't Hof C. *Helpful Hackers. How the Dutch Do Vulnerability Disclosure.* Tek Tok; 2015. https://cvth.nl/hhe.htm.

203. Koponen J, Hildén J. *Data Visualization Handbook.* Espoo, Finland: Aalto University School of Arts, Aalto ARTS books; 2019.

204. McCanless D. *Information Is Beautiful*. London: Collins; 2009.

205. McCanless D. *Knowledge Is Beautiful*. London: William Collins; 2014.

206. Lima M. *The Book of Trees. Visualizing Branches of Knowledge*. New York: Princeton Architectural Press

207. Lima, M. *The Book of Circles. Visualizing Spheres of Knowledge*. New York: Princeton Architectural Press; 2016.

208. Lima M. *Visual Complexity. Mapping Patterns of Information*. New York: Princeton Architectural Press; 2011.

209. Trend Micro. The art of cybersecurity. Trend Micro. https://www.trendmicro.com/nl_nl/config/business/campaign/art-of-cybersecurity/iframe-europe.html. Accessed November 10, 2019.

210. Trend Micro, Dawes B. *Unknown Threats Detected & Blocked Over Time*.; 2019. https://www.trendmicro.com/en_us/business/campaigns/art-of-cybersecurity.html?utm_campaign=DIG2019_Corporate_AW_CR&utm_medium=Social&utm_source=LinkedIn_Organic&utm_content=DesLauriers. Accessed November 11, 2019.

211. Trend Micro, Posavec S. *Endpoint Threats Detected & Blocked Over Time*. https://www.trendmicro.com/en_us/business/campaigns/art-of-cybersecurity.html?utm_campaign=DIG2019_Corporate_AW_CR&utm_medium=Social&utm_source=LinkedIn_Organic&utm_content=DesLauriers. Accessed November 10, 2019.

212. Parikka J. Malware as Operational Art: On the If/Then of Geopolitics and Tricksters, Lecuture presented at Het Nieuwe Instituut on 4 July 2019 on the occasion of the opening of Malware:Symptoms of Viral Infection, an exhibition about the history and evolution of the computer virus. Presented at the: The opening of Malware:Symptoms of Viral Infection, an exhibition about the history and evolution of the computer virus.; July 4, 2019; Rotterdam. https://malware.hetnieuweinstituut.nl/en/malware-operational-art-ifthen-geopolitics-and-tricksters. Accessed November 17, 2019.

213. Hyppönen M, White D. The Malware Museum. archive.org. https://archive.org/details/malwaremuseum&sort=-reviewdate. Accessed November 13, 2019.

214. Malware: Symptoms of Viral Infection, exhibition at Het Nieuwe Instituut, Rotterdam. Malware: Symptoms of Viral Infection. https://malware.hetnieuweinstituut.nl/home. Accessed November 8, 2019.

215. I love you [rev.eng]. The Aesthetics of Computer Viruses. http://www.digitalcraft.org/iloveyou/. Accessed November 17, 2019.

216. Ouchhh Studio. Poetic AI. Frame Awards. https://www.frameawards.com/winners/194254-poetic-ai. Accessed November 8, 2019.

217. Anadol R. Refik Anadol - Web page. http://refikanadol.com/. Accessed November 8, 2019.

218. Noun Project. Noun Project. https://thenounproject.com. Accessed February 7, 2019.

219. TREsPASS. Iconography principles. https://visualisation.trespass-project.eu/?p=521. Accessed March 14, 2019.

220. Roam D. *Show and Tell*. Penquin; 2014.

221. Brand W. *Visual Thinking*. Amsterdam: BIS Publishers; 2017.

Images and credits

All other images by author & Booqlab, 2020

Page number	Illustration
19	Centre for the protection of national infrastructure, UK Workmate bingo. Picture under Open Government Licence.
22	Comparison of the search interest of cyber security and information security with Google Trends. Data source: Google Trends (https://www.google.com/trends).
26	Centre for teaching quality Common Core: Paths to 21st-century success
80	Software Assurance Maturity Model OWASP
85	European Cyber Security Month Roadmap Enisa
87	Triple Bow Tie 5G Cyber Security Supervision Model Farley Wazir
89	Barriers that slow/hinder/prevent companies from managing their information as a business asset. Danette McGilvray, James Price, Tom Redman
92	Infection chain of Bad Rabbit Ransomware Trend Micro
93	Subway map to agile practices Agile Alliance
96	CyberSeek Cybersecurity career pathway CyberSeek Project and partners Burning Glass Technologies, CompTIA, National Initiative for Cybersecurity Education (NICE).
97	The Global Risks Interconnections Map 2018 World Economic Forum, design by Moritz Stefaner / Truth & Beauty

98	10 steps to cyber security National Cyber Security Centre
101	Data processing activities and compliance obligations. Pietro Calorio
102	A model for information security leadership Jeffrey Pomerantz, EDUCAUSE
104	Visualising attack graphs Concept and design: LUST (Jeroen Barendse with Eric Li) for TREsPASS Project. Data: Verizon annual Data Breach Investigations Report (DBIR) 2016
109	Occupations radar for Safety& Security. Security Talent, The Hague Security Delta (HSD)
110	Frequency of Google searchers for the words cyber security and information security in different parts of the world in different years. All maps created with Google Trends. Data source: Google Trends (https://www.google.com/trends).
112	Cyber Threat Defender Collectible Card Game The University of Texas at San Antonio's (UTSA) Center for Infrastructure Assurance and Security
113	Securing the Future State SWIFT Innotribe at Sibos 2018 Presenters: Duena Blomstrom, Jane Frankland. Drawn by Evan Wondolowski, Collective Next
114	Paper prototyping: Cloud case study TREsPASS
115	Lego simulation Northwave.(https://northwave-security.com)
118	Is your C-Suite Prepared for the Cyber Future? J. Lowe and S. Duffy, Electric Energy Online
119	Around the clock cybersecurity John McRea, Neuways
122	Data Science Hunting Funnel Austin Taylor

125	NCSC Glossary National Cyber Security Centre, UK
129	Social engineering and how to win the battle for trust (part of larger infographic). OneSpan
130	The Global Cybersecurity index 2017. International Telecommunications Union.
133	Introducing the psychology of passwords LastPass
135	Hackers Handshake photo exhibition. Photographer: Tobias Groenland (https://www.tobiasgroenland.nl).
136	Trend Micro: Unknown Threats Detected & Blocked Over Time by Brendan Dawes. Trend Micro: Endpoint Threats Detected & Blocked Over Time by Stefanie Posavec.